AF360744

Skin-Gut-Breast Microbiota Axes

Skin-Gut-Breast Microbiota Axes

Editor

Lorenzo Drago

MDPI • Basel • Beijing • Wuhan • Barcelona • Belgrade • Manchester • Tokyo • Cluj • Tianjin

Editor
Lorenzo Drago
Department of Biochemical
Sciences for Health,
University of Milan
Italy

Editorial Office
MDPI
St. Alban-Anlage 66
4052 Basel, Switzerland

This is a reprint of articles from the Special Issue published online in the open access journal *Journal of Clinical Medicine* (ISSN 2077-0383) (available at: https://www.mdpi.com/journal/jcm/special_issues/SGB_MA).

For citation purposes, cite each article independently as indicated on the article page online and as indicated below:

LastName, A.A.; LastName, B.B.; LastName, C.C. Article Title. *Journal Name* **Year**, *Volume Number*, Page Range.

ISBN 978-3-0365-0898-6 (Hbk)
ISBN 978-3-0365-0899-3 (PDF)

Contents

About the Editor

Lorenzo Drago is a professor of clinical microbiology at the University of Milan. He was the director of the laboratory medical hubs in two large Italian hospitals. He is a project leader—at the Italian and European levels—of microbiological networks focusing on microorganisms and infections, antibiotic resistance, and microbiota and probiotics. Prof. Lorenzo Drago is an international leader in laboratory diagnosis, molecular technology, next-generation sequencing systems, hospital infection control and management, antibiotic resistance, and microbiota studies. He is a co-founder of the World Association of Infection in Orthopedics and Trauma (WAIOT) and the former president of the International Society of Microbiota.

Preface to "Skin-Gut-Breast Microbiota Axes"

The "skin-gut-breast microbiota axis" comprises the network of connections—involving multiple biological systems—that allows for relational communication between the gut-skin axis, breast bacteria, and our body. This system is finely regulated, and it is crucial for maintaining the homeostasis of skin integrity, the gastrointestinal tract, and the central nervous system of humans. This network of microorganisms is known to be relevant to our health. This book describes the mechanisms of, opportunities for, and approaches to studying this system and how to harness it to improve human health.

Lorenzo Drago
Editor

Journal of
Clinical Medicine

Article

The Sulfate-Reducing Microbial Communities and Meta-Analysis of Their Occurrence during Diseases of Small–Large Intestine Axis

Ivan Kushkevych [1,*], Oľga Leščanová [1], Dani Dordevié [2], Simona Jančíková [2], Jan Hošek [3], Monika Vítězová [1], Leona Buňková [4] and Lorenzo Drago [5]

[1] Department of Experimental Biology, Faculty of Science, Masaryk University, Kamenice 753/5, 62500 Brno, Czech Republic; Lescanova@mail.muni.cz (O.L.); vitezovam@sci.muni.cz (M.V.)
[2] Department of Plant Origin Foodstuffs Hygiene and Technology, Faculty of Veterinary Hygiene and Ecology, University of Veterinary and Pharmaceutical Sciences, 61242 Brno, Czech Republic; dordevicdan@vfu.cz (D.D.); janickovasim@vfu.cz (S.J.)
[3] Regional Centre of Advanced Technologies and Materials, Faculty of Science, Palacky University in Olomouc, 78371 Olomouc, Czech Republic; jan.hosekj@upol.cz
[4] The Department of Environmental Protection Engineering, Faculty of Technology, Tomas Bata University in Zlín, 76001 Zlín, Czech Republic; bunkoval@utb.cz
[5] Department of Biomedical Sciences for Health, University of Milan, 20122 Milan, Italy; lorenzo.drago@unimi.it
* Correspondence: kushkevych@mail.muni.cz; Tel.: +420-549-495-315

Received: 28 August 2019; Accepted: 9 October 2019; Published: 11 October 2019

Abstract: Sulfate-reducing bacteria (SRB) are often isolated from animals and people with ulcerative colitis and can be involved in the IBD development in the gut–intestine axis. The background of the research consisted of obtaining mixed cultures of SRB communities from healthy mice and mice with colitis, finding variation in the distribution of their morphology, to determine pH and temperature range tolerance and their possible production of hydrogen sulfide in the small–large intestinal environment. The methods: Microscopic techniques, biochemical, microbiological, and biophysical methods, and statistical processing of the results were used. The results: Variation in the distribution of sulfate-reducing microbial communities were detected. Mixed cultures from mice with ulcerative colitis had 1.39 times higher production of H_2S in comparison with samples from healthy mice. The species of *Desulfovibrio* genus play an important role in diseases of the small–large intestine axis. Meta-analysis was also used for the observation about an SRB occurrence in healthy and not healthy individuals and the same as their metabolic processes. Conclusions: This finding is important for its possible correlation with inflammation of the intestine, where the present of SRB in high concentration plays a major part. It can be a good possible indicator of the occurrence of IBD.

Keywords: bowel disease; colitis; small–large intestine axis; sulfate reduction; hydrogen sulfide

---

1. Introduction

Sulfate-reducing bacteria (SRB) represent probably a trigger for the occurrence of inflammatory bowel diseases (IBD) since studies are connecting their presence with these diseases, especially their metabolic end product H_2S in the gut [1,2]. Other ailments (including rheumatic diseases and with ankylosing spondylitis) occur also in their presence [3]. SRB use sulfate as an electron acceptor in the process of dissimilatory sulfate reduction. The final product of this process is hydrogen sulfide [4]. Constant microorganism cultivation is happening in the large intestine since certain undigested food remains in it. [1,2]. Around 200 g of digestive material is found in the large intestine of an adult human [2,3,5,6]. These bacteria are in the fermentation process can cleave complex organic compounds and form molecular hydrogen, different acids (acetic and lactic), same as other compounds. Lactic acid bacteria fermentative properties are directly responsible for the production of lactate [4]. Other groups of microorganisms can also use lactate and acetate, serving as electron donors and carbon sources [7–12]. The important role of human physiological processes is their capability to absorb sulfate and develop amino acids out of it (cysteine and methionine). The amount of the sulfate present in the intestine is related to human diet [13–16], meaning that it is highly influenced by individual's eating habits. The importance of daily sulfate intake can be overseen by the fact that staple food commodities represent high sulfate sources (>10 μmol/g) [13].

Although, sulfate amounts that are not used in amino acid synthesis represent good conditions for SRB [1,4,17–21]. SRB needs electron acceptor (sulfate serves this purpose) and they form hydrogen sulfide as their final product [22–27]. An exogenic electron donor, including lactate can be also used and oxidized to acetate [18,28]. The dominant SRB in the intestine of humans is *Desulfovibrio* genus [5,22,28]. The studies are emphasized connections between the presence of SRB in the intestines and the prevalence of ailments, such as cholecystitis, brain abscesses, and abdominal cavity ulcerative enterocolitis. Sulfate-reducing bacteria are not the only ones that produce H_2S in the intestinal content. Numerous bacterial groups convert cysteine to H_2S, pyruvate, and ammonia by cysteine desulfhydrase activity [2–4,12].

Though connections have been found, it is still not clear how these processes are affecting the prevalence of certain ailments. Meta-analysis is used widely in medical research, as in natural science. It is included in systematic reviews as a rigorous method for mapping the evidence gained by many authors. The meta-analysis should provide unbiased overviews of multiple results and should assess evidence quality and synthesize it. The first step of a systematic review is the research question that is deconstructed by sample consideration, the second step is intervention and then come outcome and comparator. The outcome of the meta-analysis depends on the study field, but in many cases, quantitative results are used [29].

The aim of the research was to compare a variation in the morphological distribution of sulfate-reducing microbial communities from healthy mice and mice with colitis, their production of hydrogen sulfide, and to study the occurrence of these bacterial populations during diseases of the small–large intestine axis.

2. Experimental Section

2.1. Manipulation with Animals

Male C57Bl/6 mice (20 g ± 2 g) were obtained from the Animal Breeding Facility of Masaryk University (Brno, Czech Republic). They were kept under standard conditions (22 ± 2 °C, 50 ± 10% relative humidity) and alternating 12 h light/dark cycles. The animals had access to a standard diet and drinking water ad libitum. Manipulations with the animals were carried out according to the bioethical rules as per the principles of the "European Convention for the Protection of Vertebrate Animals Used for Experimental and Other Scientific Purposes" adopted in Strasbourg in 1986. The study was also approved by the "Commission for the Protection of Animals against Cruelty" and the Ethics Committee of the University of Veterinary and Pharmaceutical Sciences in Brno, Czech Republic. In total, six animals in two groups (4 + 2 animals in the first and second group, respectively) were randomly separated and used in this experiment. In the dextran sulfate sodium (DSS) group ($n = 4$), colitis was induced by administering 5% (*w/v*) DSS (MP Biomedicals, Illkirch-Graffenstaden, France, MW 36,000–50,000 Da) in drinking water for 7 days. The mice in the intact group ($n = 2$) received drinking water only. On the last day of the experiment, the animals were killed by decapitation under isoflurane anesthesia. The isolated distal colonic segments were selected for the analysis of the qualitative and quantitative composition of intestinal microflora of both groups of the animals.

2.2. Bacterial Mixed Cultures

The material used for the study consisted out of mixed sulfate-reducing bacteria cultures that were isolated from feces of healthy and with ulcerative colitis mice. After the autopsy, the samples were placed in the tubes. The bacteria were studied as mixed cultures because the aim of the study was not the purification of SRB. Mixed cultures were kept at the Laboratory of Anaerobic Microorganisms of the Department of Experimental Biology at Masaryk University (Brno, Czech Republic).

2.3. Cultivation of SRB Cultures

SRB cultures were cultivated according to Kovac and Kushkevych (2017) [30] and Postgate (1984) in a modified Postgate C medium [23]. Mohr's salt (ammonium iron sulfate hexahydrate, Sigma-Aldrich, Prague, Czech Republic) was used as a simple growth detection. Ferrous salt forms reacted with sulfide produced by SRB (dark black precipitate of FeS) and indicated the presence of SRB (the presence of dissimilatory sulfate reduction). Due to the method, it was possible to optically determine the presence of metabolic activity qualitatively and quantitatively.

The cultures were kept in medium with Mohr's salt and without is since color changes are not desirable for spectrophotometric and turbidimetric methods. In cultures kept in medium without Mohr's salt, the SRB can be detected by the sharp smell of hydrogen sulfide same as by optical turbidity. The medium was sterilized (pH 7.5–7.7, $E_h = -100$ mV). Redox potential was adjusted by Na_2S (Sigma-Aldrich, Prague, Czech Republic) and ascorbic acid (Sigma-Aldrich, Prague, Czech Republic). The anoxic atmosphere was ensured by the nitrogen gas addition, inhibiting oxygen from the air to diffuse into the medium. The oxygen proof layer was secured by the addition of paraffin (Sigma-Aldrich, Prague, Czech Republic) drops to each cultivation tube. The strains were able to grow 10 days under these conditions.

The long storage (up to one month) conditions for cultures were provided by Postgate B medium with the addition of Mohr's salt. In this medium there is always tending of bacteria to descend to the bottom of the tube due to the presence of the precipitate. Bacteria usually stick to the walls of the tube when is used modified Postgate C medium.

2.4. Description of Morphology

Microscope Olympus BX50 (lympus, Japan) was used for the observation of cells.

Phase-contrast microscopy is a technique that allows images of transparent specimens (living cells). The advantage of this technique is the possibility to do the measuring without cell killing since cells can be monitored with real-time motility. The bacterial suspension (a drop) was placed on a glass slide. The slide (cover glass added to the top of bacterial suspension) was analyzed immediately after immersion and with 100× objective.

The Gram staining method provides observation of gram-positive and gram-negative bacteria by differential staining with the use of crystal violet-iodine complex and a safranin counterstain. Gram-positive bacteria appear purple after treatment with alcohol while gram-negative bacteria appear pink. After drying samples were microscopically observed, including oil immersion 100× objective.

Capsule staining. Acidic and basic stains cannot be used for bacterial capsules. Therefore, the best way to visualize them is to stain the background using an acidic dye (e.g., nigrosine, Congo red) and to stain the cell itself using a basic stain (e.g., crystal violet, safranin, methylene blue). One drop of Congo red dye was mixed with one drop of bacterial suspension on a glass slide. After spreading throughout the slide and letting dry, it was immersed in hydrochloric acid (4 mol/L) and after a few seconds, it was let dry again. Subsequently, methylene blue dye was added on the slide and it was let standing for three minutes. After three minutes, the slide was washed with deionized water, dried, and observed with immersion oil and 100× objective. The cells were stained blue and their capsules remained white and visible on a dark background.

DAPI (4′,6-diamidino-2-phenylindole) staining is a fluorescent dye, binding by preference to the AT-rich regions of DNA [31]. Microorganisms with thick cell walls can be stained with DAPI after permeabilization of the cell wall by ethanol. For this type of microscopy, using a 48-hour old culture was found most suitable. A 48-h-old cell suspension of a volume 25 µL to 100 µL was diluted in several ml of MiliQ deionized water and washed by vacuum filtration. After washing, the filtration paper with cells was let dry. Consequently, 20 µL of DAPI stain (Sigma-Aldrich, Prague, Czech Republic) was applied and the filtration paper with cells was kept in the dark in a refrigerator for 10 min. After that, the filtration paper was washed in water, ethanol, and water, respectively, and let dry. Next, it was put on a glass slide with immersion oil applied both under and over the filtration paper with cells, and the slide was observed in a microscope, using WU filter (Sigma-Aldrich, Prague, Czech Republic) and 100× objective.

2.5. pH Tolerance and Temperature Range Test

As measured before, the optimal pH for the cultivation of intestinal SRB is from 7 to 8 [15]. The measuring was done by performing a simple pH test. The modified Postgate C medium was prepared by adjusting various pH values, performed by adding drops of sodium hydroxide (aqueous solution) and hydrochloric acid (aqueous solution), respectively. CyberScan 510 pH-meter (PreSens, Regensburg, Germany) was used to measure the exact pH values (pH ranged from 4 to 12). Media were heated to 37 °C in Wasserman tubes inoculums (obtained from healthy and not healthy mice) of cultures. Paraffin oil (500 µL) was added on the top of the medium to provide an oxygen-proof layer. The optical density of the suspension was measured at 430 nm using spectrophotometer Spectronics Genesys 5 (Thermo Fisher Scientific, Prague, Czech Republic). Blank samples were media without inoculum. Optical density was measured after 24 h of cultivation again. Bacteria were added in Eppendorf tubes and placed in thermostats (1-CUBE, Havlickuv Brod, Czech Republic) set at 5, 25, 35, 45, 50, and 60 °C. Optical density was measured at 430 nm using Spectronic Genesys 5, after 72 h of cultivation.

2.6. Production of Hydrogen Sulfide

Spectrophotometrical methylene blue method was used for measuring the presence of hydrogen sulfide in solution [32]. The bacterial suspension (1 mL) was pipetted to 5 mL of aqueous zinc acetate (5 g/L). 2 mL of p-aminodimethylaniline (Sigma-Aldrich, Prague, Czech Republic) solution (0.75 g/L in 2 M sulfuric acid) was added immediately and the solution was let stand at room temperature for 5 min. 0.5 mL of ferric chloride ($FeCl_3$) (12 g/L in 0.015 M sulfuric acid) solution was consequently added. The solution was centrifuged at 2200 RPM (10 °C for 5 min). After centrifuging, the samples lost the original light pink color and had a blue color. The absorbance was measured at 665 nm by Spectronic Genesys 5 spectrophotometer. The procedure for blank sample preparation included preparation that a clear cultivation medium was added in step 1. The concentrations used for calibration solutions ranged from 6 µmol/L to 100 µmol/L (Figure 1).

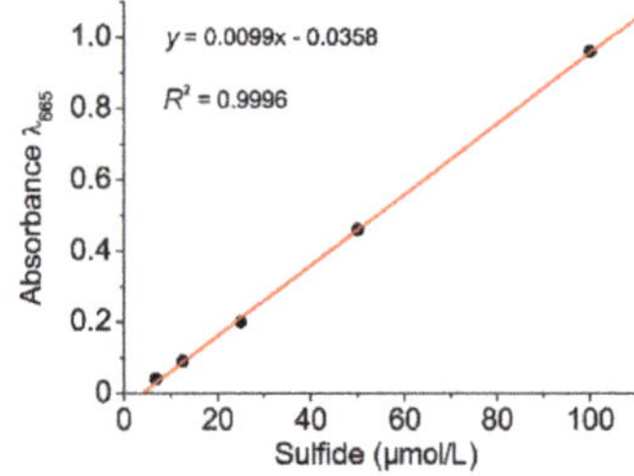

Figure 1. The calibration used for the determination of sulfide concentrations.

2.7. Statistical Analysis

Using the experimental data, the basic statistical parameters (M—mean, m—standard error, M ± m) were calculated. The accurate approximation was when $p \leq 0.0533$ [33]. Statistical analysis was done by SPSS 20 statistical software (IBM Corporation, Armonk, NY, USA). Plots were built by software package Origin 7.0 (Northampton, MA, USA).

Meta-analysis consisted of studies found on the WEB OF KNOWLEDGE database. The database found 38 studies, from the year 1945 to 2019.considering sulfate-reducing bacteria. Only six studies were included in the meta-analysis since other studies did not satisfy the specific hypothesis of the study. The Review Manager Software (Cochrane, Brno, Czech-Republic) (number 5.3 developed by Cochrane Collaboration) was used. In the included studies the data consisted of the number of participants with the positive occurrence of the SRB bacteria in the group of healthy people and people with ulcerative colitis. In other studies, the data consisted of the mean, standard deviation and the number of the measurements. Heterogeneity was expressed by the I^2 test, where the higher I^2 represented a higher heterogeneity.

3. Results

The vibrio shape was a dominant shape of the cells, as expected. Though they are very small and thin that makes them very often hard to be observed. These cells were marked as *Desulfovibrio* sp. Due to their characteristic shape, gram negativity and flagellar motility (Figure 2). Very abundant were also cells, oval form. Chain and cluster shaped had cocci that were larger than vibrios, same as some rod shape cells were observed too. Rods have almost similar characteristics as cocci. Not abundantly spirilloid forms of bacteria were present too. They had long shape and were very thin, curved multiple times (maximum twelve curves) (Figure 2A). They had long, polar flagella that are responsible for rapid movement. Gram-negative bacteria only were not only present in SRB cultures isolated from rodents (Figure 2B).

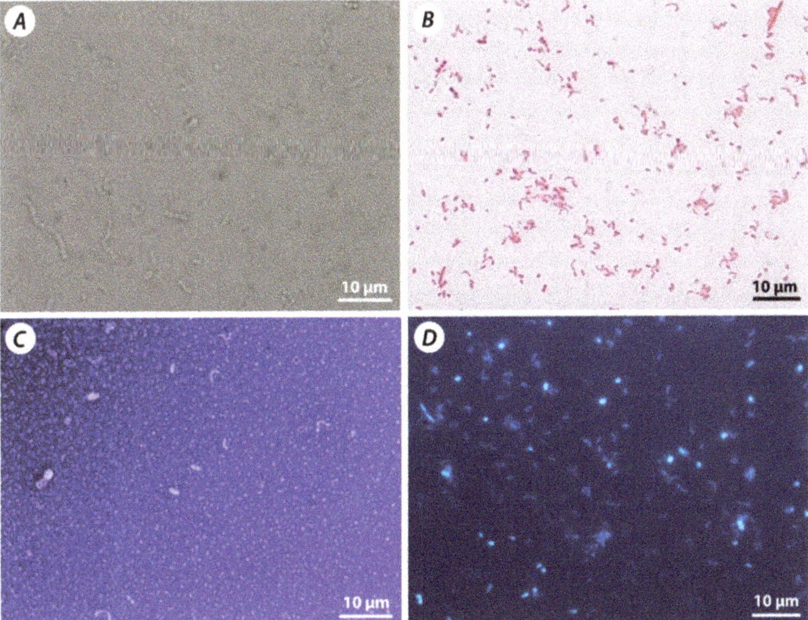

Figure 2. Sulfate-reducing bacteria (SRB) mixed culture: native slide (**A**), Gram staining (**B**), capsule staining (**C**), DAPI staining (**D**).

Desulfotomaculum is rod-shaped (stained Gram-positive) (representing non-SRB genera in the gut) can be seen in Figure 2C since it has a short rod oval shape. According to the previous microscopic technique, cocci can be encapsulated or not. More often encapsulated cocci are present in pairs. The formation of capsules occurs probably due to a non-favorable environment, such as high hydrogen sulfide concentrations due to sulfate-reducing bacteria presence. It is important to stress out that capsule formation is not defined as SRB characteristic. DAPI (4′,6-diamidino-2-phenylindole) staining is compliant with the observations made by the previous technique (Figure 2D). The most abundant was vibrio cell-shape. SRB present in the gut isolate was probably *Desulfovibrio* sp., according to literature data that is describing them as the most frequently isolated species in the intestinal inflammation environment. Cocci were confirmed by DAPI staining since they are significantly brighter and larger than other cells. The findings that DAPI cultures bind to DNA molecules indicate that some oval-shaped have more DNA than others, meaning that they are unrelated to each other. Different sizes of cocci, gained by previous techniques, is supporting this interpretation. These cells were found in multiple isolates because thin rods of exceeding length were found by DAPI staining. These cells represent a common microbiome in the intestines that are capable to survive in conditions designed for SRB cultivation.

The fastest bacterial growth and viability, measured spectrophotometrically OD_{430} (Figure 3), was detected after 24 h of cultivation at 37 °C and pH from 8.0 to 9.0. A significant drop in viability was observed at pH 10. The absence of black precipitate was observed in tubes with Mohr's salt and pH > 10 (Figure 3A). This result is indicating a threshold limit pH ≥ 10 both for sulfate-reducers and other (contaminating) species. The values did not reach zero value but were stabilized at around 30–40% of maximum bacterial growth. It means that bacteria were capable to survive and divide at this pH, reaching an optical density of 0.3. Black precipitate occurred at all pH values, meaning that bacteria can survive a longer time period before starting to metabolize and produce hydrogen sulfide. The changing of color in the tubes at pH 11 and 12 occurred due to basic conditions. It means that the measured values of optical density can be explained by the extreme pH effect.

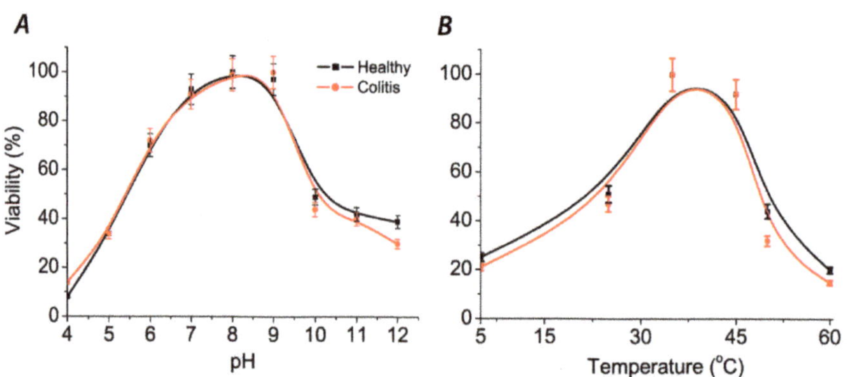

Figure 3. Various pH (**A**) and temperature (**B**) influence on relative viability of SRB cultures.

After 72 h of cultivation bacterial growth of all samples was observed. SRB cultures can grow at various ranges of temperature conditions, not only at 37 °C, though the fastest growth occurred at temperature ranges from 37 °C to 45 °C. Another observation was that cells survived for three days at 50 °C and died on the temperatures higher than 60 °C and at the temperature of 5 °C (no bacterial growth, no hydrogen sulfide production, black precipitate not occurred and low OD_{430} values were measured. The growth was slow at a temperature of 25 °C. The relative viability values of SRB are shown in Figure 3B.

The concentrations of H_2S in time change according to cell number, same as their metabolic activity rate. The maximum measured hydrogen sulfide concentrations were measured after 48 h of cultivation

(Figure 4). After 48 h of cultivation H_2S concentrations decreased due to the decrease in relative substrate concentration in the medium, though H_2S can clear out from the medium. H_2S is present in a soluble form in the medium and can be released as the gaseous phase (the presence of a bubble under the lid, accompanied by hydrogen sulfide sharp smell) into the environment. Consequently, sulfide concentrations dropped at the beginning of the cultivation. After six hours of cultivation, soluble sulfide was eliminated into gaseous phase and it was a point where the lowest H_2S levels were detected. Mixed cultures from mice with ulcerative colitis had 1.39 times higher production of H_2S in comparison with samples from healthy mice. The maximal difference was 20 μmol/L after 48 h of cultivation.

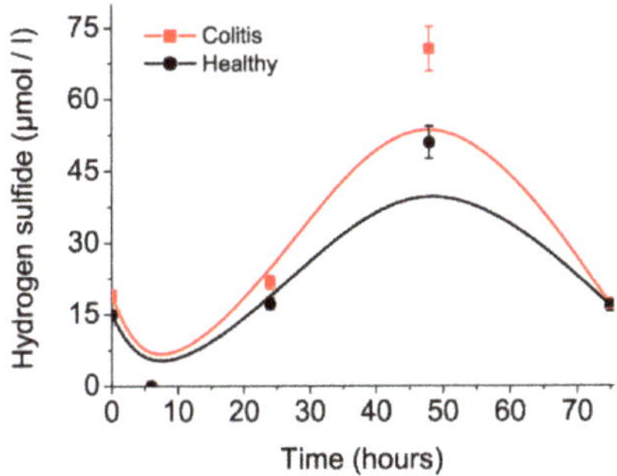

Figure 4. Amount of hydrogen sulfide in cultivation medium in 72 h.

It should be noted that sulfate-reducing microbial communities from healthy mice and mice with colitis were used only as of the model objects for confirmation of morphology distribution and hydrogen sulfide production in different groups of animals (healthy and with ulcerative colitis). Another part of the study consisted of a literature data overview that was conducted by meta-analysis. This method was used for comparing SRB prevalence in healthy individuals and people with developed inflammatory bowel disease. The occurrence of SRB in a group of healthy people and patients with ulcerative colitis (UC) was studied (Figure 5). The location of the square on the right side means that not healthy people are more likely to experience SRB. A significant difference in the occurrence of SRB in healthy people can be observed in the first study [34]. The other two studies [5,35] already touch the zero effect line at a 95% confidence interval, so there is no significant difference. The diamond can then be seen on the right side. Summary of the studies found that SRB is less common in healthy people than in people with UC.

Study or Subgroup	UC Events	Total	Healthy Events	Total	Weight	Risk Ratio M-H, Fixed, 95% CI	Risk Ratio M-H, Fixed, 95% CI
Coutinho et al. 2017	17	21	5	24	33.7%	3.89 [1.73, 8.71]	
Loubinoux et al. 2002	3	5	10	41	15.7%	2.46 [1.00, 6.03]	
Zinkevich and Beech 2000	12	13	7	13	50.6%	1.71 [1.01, 2.90]	
Total (95% CI)		39		78	100.0%	2.56 [1.68, 3.91]	
Total events	32		22				
Heterogeneity: Chi² = 3.27, df = 2 (P = 0.20); I² = 39%							
Test for overall effect: Z = 4.36 (P < 0.0001)							

Figure 5. The occurrence of SRB in a group of healthy people and patients with UC.

The production of hydrogen sulfide occurs in the process of dissimilatory sulfate reduction, where tree main enzymes are involved. Since the species of *Desulfovibrio* genus were dominant among SRB in both mice and people with ulcerative colitis, the activity of the enzymes involved in the processes of sulfate reduction in *Desulfovibrio* and other intestinal SRB *Desulfomicrobium* was compared (Figure 6). In the case of enzyme activity in cell-free extracts, it was found that in all cases it had the lower enzymatic activity of *Desulfomicrobium* sp. phosphotransacetylase and pyruvate-ferredoxin activity was more or less the same in *Desulfovibrio* bacteria. Thus, it can be argued that the activity of Na$^+$/K$^+$ ATPase is the highest of the investigated enzymes in the cell-free extracts of *Desulfovibrio*. Similar results were observed in soluble fractions. The activity of Na$^+$/K$^+$ ATPase is highest in *Desulfovibrio* than *Desulfomicrobium* in all enzymes examined. In the case of sediment fractions, higher Na$^+$/K$^+$ ATPase activity was again found in *Desulfovibrio* bacteria and no activity was observed in both *Desulfovibrio* and *Desulfomicrobium* in the other investigated enzymes, phosphotransacetylase, and pyruvate-ferredoxin oxidoreductase.

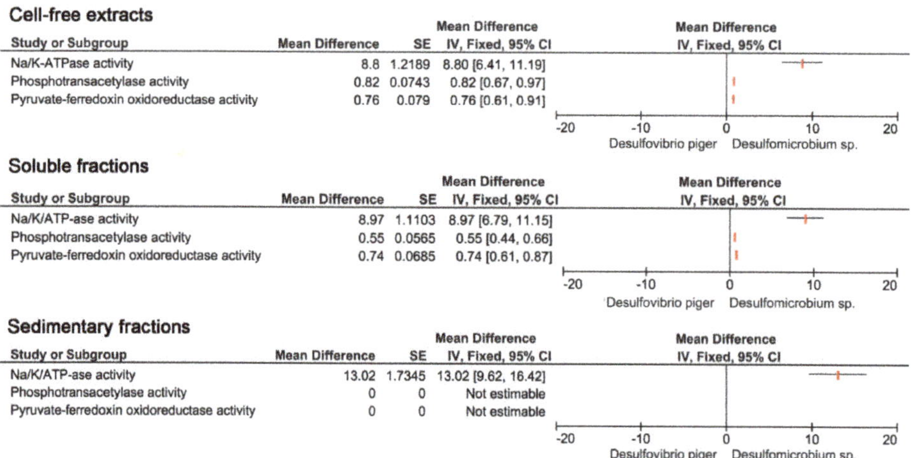

Figure 6. Enzyme activity in *Desulfovibrio* and *Desulfomicrobium*.

Thus, the contribution of sulfate-reducing microbial communities, especially of the *Desulfovibrio* genus, in both groups of healthy people and patients with UC and enzymatic activities of bacterial cells is based on a meta-analysis is obvious. Though, the number of studies is certainly not enough for a stronger conclusion.

4. Discussion

Important factors that influence the intestinal environment are sulfate consumption, sulfide production, lactate consumption and acetate accumulation [7–10]. Very often Desulfovibrio genus is present in the intestines and feces of people and animals with inflammatory bowel disease, meaning that this genus plays an important role in the development and occurrence of this ailment. Sulfate is used as a terminal electron acceptor by these bacteria, the same as organic compounds are used as electron donors in their metabolism [6,7]. Leading us to the conclusion that sulfate in food commodities (some bread, soya flour, dried fruits, brassicas, and sausages, as well as some beers, ciders, and wines) play an important role in the development of bowel disease [13].

The principal component analysis showed that the *Desulfovibrio* strains from individuals with colitis grouped in one cluster by biomass accumulation and sulfide production, while the strains from healthy individuals formed another cluster that included the same parameters. A negative correlation (Pearson correlations, $p < 0.01$) was found between sulfate and lactate consumption. Biomass accumulation and hydrogen sulfide showed lower linear regression (R^2). The kinetic parameters, biomass accumulation, and sulfide production have an important role in bowel inflammation, including ulcerative colitis. Acetate produced by SRB probably has a synergy interaction with H$_2$S since sulfate consumption and lactate oxidation represent minor factors in bowel disease [16].

Optimum growing conditions for the bacteria were provided by the study. The intensive growth of *D. piger* Vib-7 was observed in the presence of higher electron acceptor and donor concentrations.

Consequently, the intensive accumulation of sulfide and acetate occurs too. According to previous studies and literature data, these conditions are the probable cause of ulcerative colitis, leading to bowel cancer. Hydrogen sulfide negatively affects intestinal mucosa, epithelial cells, the growth of colonocytes [4,14–18,36–39], causes phagocytosis, causes the death of intestinal bacteria [4,12,24], and induces hyperproliferation and metabolic abnormalities of epithelial cells [12]. The presence of SRB and high level of metabolites are also connected with colon inflammation [4,6,38]. Hydrogen sulfide concentrations are regulating the integrity of colonocytes [37–39]. In the samples of individuals with ulcerative colitis was also found that SRB sulfide production is higher [5,6]. According to another study dealing with the SRB metabolic process was found that the strains isolated from people with colitis shifted to the right side of the Y-axis by biomass accumulation, sulfate consumption, lactate oxidation, same as hydrogen sulfide and acetate production, in comparison with the strains isolated from healthy individuals. The percentages were differences observed in shifting to the right side of the Y-axis: biomass accumulation 26%, sulfate consumption 1.5%, and sulfide production 5% [14]. The intestinal microbiota is a complex system, interactions occur between clostridia, methanogens, lactic acid bacteria, etc. Though, SRB plays a central role in the development of IBD, including ulcerative colitis [1–3,11]. Lactic acid bacteria, methanogens, and many other intestinal microorganisms can be inhibited by hydrogen sulfide produced by SRB [2].

Preservatives added to food often contain sulfur oxides, sulfate polysaccharides (mucin), chondroitin sulfate, carrageenan, and other food commodities represent the source of sulfate and lead to evaluated sulfate intake in the daily diet that leads to increase of hydrogen sulfide concentrations produced by SRB. The western diet contains over 16.6 mmol sulfate/day [13] and the feces of about 50% of healthy individuals contain SRB (*Desulfovibrio*: up to 92%) [1,5,24]. On the other hand, the concentrations of hydrogen sulfide are toxic not only for the intestinal environment but also for their producers. The concentrations higher than 6 mM stop the growth of *Desulfovibrio*, but metabolic activity was not 100% inhibited (the results supported by cross-correlation and principal component analysis). 5 mM concentrations of H_2S resulted in two times and eight times longer lag phase and generation time, respectively [18]. It should be noted that clostridia can also produce hydrogen sulfide, but in smaller quantities and can be interacted with SRB [40] Terminal oxidative processes in the large intestine of humans can be also included in the activities of SRB. The connections between SRB presence and activity in the intestine and occurrence of ulcerative colitis were also found in animal studies where SRB isolated from mice with UC produced 1.14 times (higher hydrogen sulfide production rate can damage aggressively intestinal mucosa) more sulfide ions than SRB isolates from healthy mice [6].

It is of crucial importance that all issues concerning H_2S metabolic processes and its influence on the gastrointestinal environment are well studied and tested. Since it has been observed in animal studies that H_2S-releasing agents can be seen as promising therapeutic agents for many indications [41]. H_2S is confirmed to represent an important signaling factor for cardiovascular and nervous systems statute [42]. The way how cecal musoca protects itself from the toxical effects of H_2S is the conversion to thiosulfate. Consequently, these metabolic pathways play an important role in the occurrence of ulcerative colitis [43]. The importance of similar studies can be seen through the fact that mechanisms leading to Chron's disease still remain unclear [44].

According to meta-analysis, SRB occurs more often in patients with UC. The finding can be explained by the fact that counts of SRB are lower (though still detectable) in healthy individuals. Oppositely, in patients with developed inflammatory bowel disease, the production of H_2S reaches toxic levels and also destroyed its producers (sulfate-reducing bacteria) [15].

5. Conclusions

Sulfate-reducing bacteria are present in various environments and they make a high impact on animal and human health since their presence is a possible contributing factor in the development of inflammatory bowel diseases. Their morphology (vibrio, spiral, rods, and cocci) and diversity are highly influenced by environmental conditions including temperature, pH, oxygen presence and substrate availability. Unique in nature is anaerobic sulfate-reducing bacteria metabolism in which hydrogen sulfide is produced in the process of electron acceptors (mainly sulfate ions) reduction (the process of dissimilatory sulfate reduction). The study clearly showed that mixed SRB cultures obtained from healthy and with ulcerative mice were equally polymorphic (the most often vibrio and coccus shape occurred). Though, the production of hydrogen sulfide differs significantly among isolated cultures. It was observed that isolates from not healthy mice produced higher hydrogen sulfide amounts. This observation is emphasizing correlations between intestine inflammation occurrence and hydrogen sulfide concentrations. The meta-analysis confirmed these correlations. Presently, it is still not fully understood the occurrence processes of inflammatory bowel diseases, including ulcerative colitis. Though, the study is emphasizing one more time that the occurrence of SRB in the samples with developed IBD is pointing out the importance of issues concerning sulfate-reducing bacteria.

Author Contributions: Conceptualization, I.K., O.L., and D.D.; methodology, I.K., O.L., and M.V.; validation, S.J., M.V., and D.D.; formal analysis, S.J., M.V., and L.D.; investigation, O.L., I.K.; resources, I.K.; data curation, L.B., D.D.; writing—original draft preparation, I.K., D.D., S.J., and M.V.; writing—review and editing, I.K., J.H., L.B., and L.D.; visualization, I.K.; supervision, M.V.; project administration, I.K.; funding acquisition, I.K., D.D., and M.V.

Funding: This research was supported by Grant Agency of the Masaryk University (MUNI/A/0902/2018).

Conflicts of Interest: The authors declare no conflict of interest.

References

1. Gibson, G.R.; Cummings, J.H.; Macfarlane, G.T. Growth and activities of sulphate-reducing bacteria in gut contents of health subjects and patients with ulcerative colitis. *FEMS Microbiol. Ecol.* **1991**, *86*, 103–112. [CrossRef]

2. Gibson, G.R.; Macfarlane, S.; Macfarlane, G.T. Metabolic interactions involving sulphate-reducing and methanogenic bacteria in the human large intestine. *FEMS Microbiol. Ecol.* **1993**, *12*, 117–125. [CrossRef]

3. Cummings, J.H.; Macfarlane, G.T.; Macfarlane, S. Intestinal Bacteria and Ulcerative Colitis. *Curr. Issues Intest. Microbiol.* **2003**, *4*, 9–20. [PubMed]

4. Barton, L.L.; Hamilton, W.A. *Sulphate-Reducing Bacteria Environmental and Engineered Systems*; Cambridge University Press: Cambridge, UK, 2017.

5. Loubinoux, J.; Bronowicji, J.P.; Pereira, I.A. Sulphate-reducing bacteria in human feces and their association with inflammatory diseases. *FEMS Microbiol. Ecol.* **2002**, *40*, 107–112. [CrossRef] [PubMed]

6. Kováč, J.; Vítězová, M.; Kushkevych, I. Metabolic activity of sulfate-reducing bacteria from rodents with colitis. *Open Med.* **2018**, *13*, 344–349. [CrossRef] [PubMed]

7. Kushkevych, I.; Vítězová, M.; Fedrová, P.; Vochyanová, Z.; Paráková, L.; Hošek, J. Kinetic properties of growth of intestinal sulphate-reducing bacteria isolated from healthy mice and mice with ulcerative colitis. *Acta Vet. Brno* **2017**, *86*, 405–411. [CrossRef]

8. Kushkevych, I.; Fafula, R.; Parak, T.; Bartoš, M. Activity of Na^+/K^+-activated Mg^{2+}-dependent ATP hydrolase in the cell-free extracts of the sulfate-reducing bacteria *Desulfovibrio piger* Vib-7 and *Desulfomicrobium* sp. Rod-9. *Acta Vet. Brno* **2015**, *84*, 3–12. [CrossRef]

9. Kushkevych, I.V. Activity and kinetic properties of phosphotransacetylase from intestinal sulfate-reducing bacteria. *Acta Biochem. Pol.* **2015**, *62*, 1037–1108. [CrossRef]

10. Kushkevych, I.V. Kinetic Properties of Pyruvate Ferredoxin Oxidoreductase of Intestinal Sulfate-Reducing Bacteria *Desulfovibrio piger* Vib-7 and *Desulfomicrobium* sp. Rod-9. *Pol. J. Microbiol.* **2015**, *64*, 107–114.

11. Loubinoux, J.; Mory, F.; Pereira, I.A.; Le Faou, A.E. Bacteremia caused by a strain of *Desulfovibrio* related to the provisionally named *Desulfovibrio fairfieldensis*. *J. Clin. Microbiol.* **2000**, *38*, 931–934.

12. Pitcher, M.C.; Cummings, J.H. Hydrogen sulphide: A bacterial toxin in ulcerative colitis? *Gut* **1996**, *39*, 1–4. [CrossRef] [PubMed]

13. Florin, T.H.; Neale, G.; Goretski, S. Sulfate in food and beverages. *J. Food Compos. Anal.* **1993**, *6*, 140–151. [CrossRef]

14. Kushkevych, I.; Dordević, D.; Vítězová, M.; Kollár, P. Cross-correlation analysis of the *Desulfovibrio* growth parameters of intestinal species isolated from people with colitis. *Biologia* **2018**, *73*, 1137–1143. [CrossRef]

15. Kushkevych, I.; Dordević, D.; Vítězová, M. Analysis of pH dose-dependent growth of sulfate-reducing bacteria. *Open Med.* **2019**, *14*, 66–74. [CrossRef] [PubMed]

16. Kushkevych, I.; Dordević, D.; Kollar, P. Analysis of physiological parameters of *Desulfovibrio* strains from individuals with colitis. *Open Life Sci.* **2018**, *13*, 481–488. [CrossRef]

17. Kushkevych, I.; Vítězová, M.; Kos, J.; Kollár, P.; Jampilek, J. Effect of selected 8-hydroxyquinoline-2-carboxanilides on viability and sulfate metabolism of *Desulfovibrio piger*. *J. Appl. Biomed.* **2018**, *16*, 241–246. [CrossRef]

18. Kushkevych, I.; Dordević, D.; Vítězová, M. Toxicity of hydrogen sulfide toward sulfate-reducing bacteria *Desulfovibrio piger* Vib-7. *Arch. Microbiol.* **2019**, *201*, 389–397. [CrossRef] [PubMed]

19. Kushkevych, I.; Kollar, P.; Suchy, P.; Parak, T.; Pauk, K.; Imramovsky, A. Activity of selected salicylamides against intestinal sulfate-reducing bacteria. *Neuroendocrinol. Lett.* **2015**, *36*, 106–113. [PubMed]

20. Kushkevych, I.; Kollar, P.; Ferreira, A.L.; Palma, D.; Duarte, A.; Lopes, M.M.; Bartos, M.; Pauk, K.; Imramovsky, A.; Jampilek, J. Antimicrobial effect of salicylamide derivatives against intestinal sulfate-reducing bacteria. *J. Appl. Biomed.* **2016**, *14*, 125–130. [CrossRef]

21. Kushkevych, I.; Kos, J.; Kollar, P.; Kralova, K.; Jampilek, J. Activity of ring-substituted 8-hydroxyquinoline-2-carboxanilides against intestinal sulfate-reducing bacteria *Desulfovibrio piger*. *Med. Chem. Res.* **2018**, *27*, 278–284. [CrossRef]

22. Loubinoux, J.; Valente, F.M.A.; Pereira, I.A.C. Reclassification of the only species of the genus *Desulfomonas*, *Desulfomonas pigra*, as *Desulfovibrio piger* comb. nov. *Int. J. Syst. Evol. Microbiol.* **2002**, *52*, 1305–1308. [PubMed]

23. Postgate, J.R. *The Sulfate Reducing Bacteria*; Cambridge University Press: Cambridge, UK, 1984.

24. Rowan, F.E.; Docherty, N.G.; Coffey, J.C.; O'Connell, P.R. Sulphate-reducing bacteria and hydrogen sulphide in the aetiology of ulcerative colitis. *Br. J. Surg.* **2009**, *96*, 151–158. [CrossRef] [PubMed]

25. Kushkevych, I.; Vítězová, M.; Vítěz, T.; Bartoš, M. Production of biogas: Relationship between methanogenic and sulfate-reducing microorganisms. *Open Life Sci.* **2017**, *12*, 82–91.

26. Kushkevych, I.; Vítězová, M.; Vítěz, T.; Kovac, J.; Kaucká, P.; Jesionek, W.; Bartoš, M.; Barton, L. A new combination of substrates: Biogas production and diversity of the methanogenic microorganisms. *Open Life Sci.* **2018**, *13*, 119–128. [CrossRef]

27. Kushkevych, I.; Kováč, J.; Vítězová, M.; Vítěz, T.; Bartoš, M. The diversity of sulfate-reducing bacteria in the seven bioreactors. *Arch. Microbiol.* **2018**, *200*, 945–950. [CrossRef] [PubMed]

28. Kushkevych, I.; Dordević, D.; Kollar, P.; Vítězová, M.; Drago, L. Hydrogen Sulfide as a Toxic Product in the Small–Large Intestine Axis and its Role in IBD Development. *J. Clin. Med.* **2019**, *8*, 1054. [CrossRef] [PubMed]

29. Mallett, R.; Hagen Zanker, J.; Slater, R.; Duvendack, M. The benefits and challenges of using systematic reviews in international development research. *J. Dev. Eff.* **2012**, *4*, 445–455. [CrossRef]

30. Kováč, J.; Kushkevych, I. New modification of cultivation medium for isolation and growth of intestinal sulfate-reducing bacteria. In Proceedings of the International PhD Students Conference MendelNet, Brno, Czech Republic, 6–7 November 2019; pp. 702–707.

31. Stan-Lotter, H.; Leuko, S.; Legat, A.; Fendrihan, S. The Assessment of the Viability of Halophilic Microorganisms in Natural Communities. *Methods Microbiol.* **2006**, *35*, 569–584.

32. Cline, J.D. Spectrophotometric determination of hydrogen sulfide in natural water. *Limnol. Oceanogr.* **1969**, *14*, 454–458. [CrossRef]

33. Bailey, N.T.J. *Statistical Methods in Biology*; Cambridge University Press: Cambridge, UK, 1995.

34. Coutinho, C.M.L.M.; Coutinho-Silva, R.; Zinkevich, V.; Pearce, C.B.; Ojcius, D.M.; Beech, I. Sulphate-reducing bacteria from ulcerative colitis patients induce apoptosis of gastrointestinal epithelial cells. *Microb. Pathog.* **2017**, *112*, 126–134. [CrossRef]

35. Zinkevich, V.V.; Beech, I.B. Screening of sulfate-reducing bacteria in colonoscopy samples from healthy and colitic human gut mucosa. *FEMS Microbiol. Ecol.* **2000**, *34*, 147–155. [CrossRef] [PubMed]

36. Attene-Ramos, M.S.; Wagner, E.D.; Plewa, M.J.; Gaskins, H.R. Evidence that hydrogen sulfide is a genotoxic agent. *Mol. Cancer Res.* **2006**, *4*, 9–14. [CrossRef] [PubMed]

37. Beauchamp, R.O.; Bus, J.S.; Popp, J.A.; Boreiko, C.J.; Andjelkovich, D.A.; Leber, P. A critical review of the literature on hydrogen sulfide toxicity. *CRC Crit. Rev. Toxicol.* **1984**, *13*, 25–97. [CrossRef] [PubMed]

38. Blachier, F.; Davila, A.M.; Mimoun, S. Luminal sulfide and large intestine mucosa: Friend or foe? *Amino Acids* **2010**, *39*, 335–347. [CrossRef] [PubMed]

39. Grieshaber, M.K.; Völkel, S. Animal adaptations for tolerance and exploitation of poisonous sulfide. *Annu. Rev. Physiol.* **1998**, *60*, 33–53. [CrossRef] [PubMed]

40. Černý, M.; Vítězová, M.; Vítěz, T.; Bartoš, M.; Kushkevych, I. Variation in the Distribution of Hydrogen Producers from the Clostridiales Order in Biogas Reactors Depending on Different Input Substrates. *Energies* **2018**, *11*, 3270. [CrossRef]

41. Wallace, J.L.; Ferraz, J.G.; Muscara, M.N. Hydrogen sulfide: An endogenous mediator of resolution of inflammation and injury. *Antioxid. Redox Signal.* **2012**, *17*, 58–67. [CrossRef] [PubMed]

42. Szabó, C. Hydrogen sulphide and its therapeutic potential. *Nat. Rev. Drug Discov.* **2007**, *6*, 917–935. [CrossRef]

43. Levitt, M.D.; Furne, J.; Springfield, J.; Suarez, F.; DeMaster, E. Detoxification of hydrogen sulfide and methanethiol in the cecal mucosa. *J. Clin. Investing.* **1999**, *104*, 1107–1114. [CrossRef]

44. Mottawea, W.; Chiang, C.K.; Mühlbauer, M.; Starr, A.E.; Butcher, J.; Abujamel, T.; Hajibabaei, M. Altered intestinal microbiota–host mitochondria crosstalk in new onset Crohn's disease. *Nat. Commun.* **2016**, *7*, 13419. [CrossRef]

Journal of
Clinical Medicine

Article

Analysis of Gut Microbiota and Their Metabolic Potential in Patients with Schizophrenia Treated with Olanzapine: Results from a Six-Week Observational Prospective Cohort Study

Justyna Pełka-Wysiecka [1], Mariusz Kaczmarczyk [2], Agata Bąba-Kubiś [1], Paweł Liśkiewicz [1], Michał Wroński [1], Karolina Skonieczna-Żydecka [3], Wojciech Marlicz [4], Błażej Misiak [5], Teresa Starzyńska [4], Jolanta Kucharska-Mazur [1], Igor Łoniewski [1,*] and Jerzy Samochowiec [1]

[1]   Department of Psychiatry, Pomeranian Medical University in Szczecin, Broniewskiego 26, 71-460 Szczecin, Poland; wysiecki@wp.pl (J.P.-W.); agata0621@gmail.com (A.B.-K.); pjliskiewicz@gmail.com (P.L.); mwronski@pum.edu.pl (M.W.); jola_kucharska@tlen.pl (J.K.-M.); samoj@pum.edu.pl (J.S.)
[2]   Department of Clinical and Molecular Biochemistry, Pomeranian Medical University in Szczecin, Powstańców Wielkopolskich 72, 70-111 Szczecin, Poland; mariush@pum.edu.pl
[3]   Department of Human Nutrition and Metabolomics, Pomeranian Medical University in Szczecin, Broniewskiego 24, 71-460 Szczecin, Poland; karzyd@pum.edu.pl
[4]   Department of Gastroenterology, Pomeranian Medical University in Szczecin, Unii Lubelskiej 1, 71-252 Szczecin, Poland; marlicz@hotmail.com (W.M.); testa@pum.edu.pl (T.S.)
[5]   Department of Genetics, Wroclaw Medical University, Marcinkowskiego 1, 50-368 Wrocław, Poland; mblazej@interia.eu
*   Correspondence: sanprobi@sanprobi.pl; Tel.: +48-91-441-4806

Received: 13 August 2019; Accepted: 25 September 2019; Published: 3 October 2019

Abstract: Accumulating evidence indicates the potential effect of microbiota on the pathogenesis and course of schizophrenia. However, the effects of olanzapine, second-generation antipsychotics, on gut microbiota have not been investigated in humans. This study aimed to analyze fecal microbiota in schizophrenia patients treated with olanzapine during six weeks of their hospital stay. After a seven-day washout from all psychotropic medications, microbiota compositions were evaluated at baseline and after six weeks of hospitalization using 16S rRNA sequencing. The study was conducted in 20 inpatients, who followed the same hospital routine and received 5–20 mg daily doses of olanzapine. Olanzapine treatment was associated with clinical improvements in all patients and significant increases in body mass index in females, but not changes in gut microbiota compositions and predicted function. The severity of symptoms at the beginning of treatment varied in accordance with the predicted metabolic activity of the bacteria. The present findings indicate that the microbiota of schizophrenia patients is highly individual and has different taxonomical (Type 1, with a predominance of *Prevotella*, and Type 2 with a higher abundance of *Bacteroides*, *Blautia* and *Clostridium*) and functional clusters, and it does not change following six weeks of olanzapine therapy; in addition, the microbiota is not associated with either the weight gain observed in women or the effectiveness of olanzapine therapy.

Keywords: microbiota; schizophrenia; olanzapine administration; weight gain

---

1. Introduction

More than 21 million people worldwide suffer from schizophrenia (SZ) [1]. A growing body of studies has shown the role of the gut–brain axis dysregulation in the pathophysiology of SZ. Subclinical inflammation, aberrant monoamine metabolism, and abnormal hypothalamic–pituitary–adrenal axis activation have been widely reported in patients with SZ [2–5] and are associated with microbiota alterations [6–9]. For instance, Schwartz et al. [10] found elevated abundance of Lactobacillaceae, Halothiobacillaceae, Brucellaceae, and Micrococcineae and lowered counts of Veillonellaceae in a cohort of SZ patients; in addition, greater microbial abnormalities, lower remission rates, and poorer responses to therapy, as well as decreased microbiome α-diversity index and altered gut microbial composition, were observed in SZ patients [11]. Although mechanisms underlying the potential effect of microbiota on the pathogenesis and course of SZ are yet to be determined, chronic inflammation [12] and altered tryptophan metabolism [13,14] have been suggested to be implicated in the pathogenesis of SZ. However, gut microbiota-associated biomarkers that would hold clinical utility have not been indicated to date.

Olanzapine (OLZ), one of the most widely used second-generation antipsychotics (SGAs) [15], has multiple adverse effects, including weight gain, dyslipidemia, impaired glucose metabolism, and hypertension [16–19]. These metabolic adversities may occur shortly after treatment implementation and progress with treatment duration [20–22]. Importantly, the first year of antipsychotic treatment is a critical period for weight gain and other metabolic adverse effects [23]. Notably, weight gain at the beginning of OLZ therapy can be used to predict long-term outcomes related to cardiovascular comorbidity. Therefore, dietary counseling and weight management, including regular bodyweight measurements, should be implemented as soon as the OLZ therapy begins [24,25]. However, weight gain is of multifactorial nature [20,26–28], and, to date, no effective therapeutic strategies could prevent weight gain in patients treated with OLZ.

A few studies have demonstrated that OLZ administration plays a role in weight gain and metabolic malfunctions. Davey et al. [29] found that OLZ treatment induced metabolic alterations via microbiota changes, and the metabolic alterations could be reversed by treatment with antibiotics; in addition, microbial, inflammatory, and metabolic adversities related to OLZ treatment were sex-dependent [30]. Moreover, Morgan et al. [31] observed that weight gain depended on gut microbiota, and specific bacteria were responsible for weight gain. Furthermore, Flowers et al. [32] revealed that clusters of gut microbiota were associated with pharmacological treatment in patients with bipolar disorder. However, to the best of our knowledge, the effects of OLZ on gut microbiota in patients with SZ have not been investigated. We hypothesized that short-term treatment with OLZ in controlled conditions (unified dietary intake and environmental factors) affects fecal microbiota compositions, and microbiota can affect body weight and treatment efficacy. Accordingly, this study analyzed microbiota compositions of stool samples collected from a cohort of SZ inpatients. The cohort comprised of acutely-relapsed SZ inpatients who were followed-up for six weeks during OLZ treatment.

2. Materials and Methods

2.1. Patients

The study protocol was approved by the Bioethics Committee of the Pomeranian Medical University in Szczecin (Poland). All participants received a written description of the study aims and provided written informed consent before participation. Participants were recruited as inpatients at the Department of Psychiatry in Szczecin (Poland) between October 2016 and May 2018, and only 20 psychiatric inpatients met the inclusion criteria. The flow chart of the study design is shown in Figure 1. SZ was diagnosed based on the ICD (International Statistical Classification of Diseases and Related Health Problems) −10 criteria.

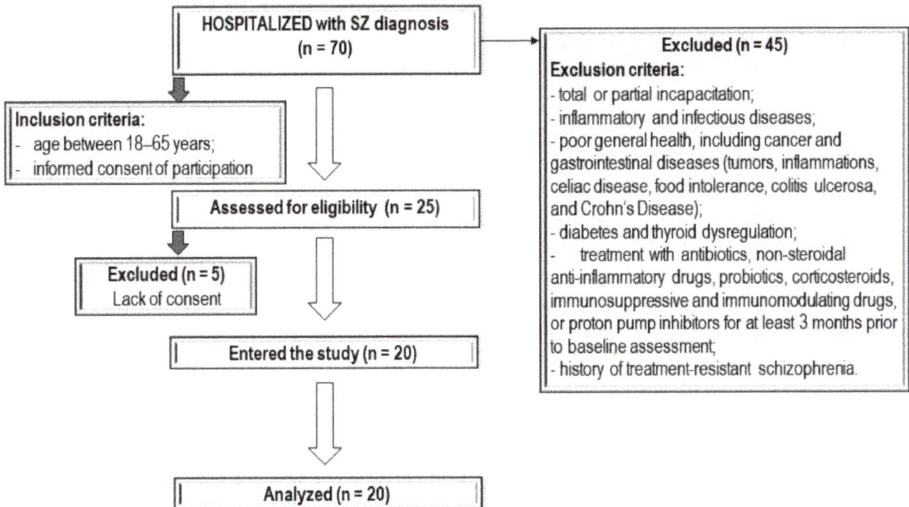

Figure 1. Flow chart of the study design. SZ, schizophrenia.

2.2. Study Protocol

All participants were subjected to the same daily activities, including physical exercise (daily morning exercise and a walk with a therapist), occupational therapy, and psycho-educational activities. Two senior psychiatrists performed the psychiatric and basic physical examinations, and a gastroenterologist conducted a comprehensive physical examination.

Patients received a standard hospital diet (i.e., 2995 ± 93 kcal, 106 ± 14 g total protein, 420 ± 24 g carbohydrates, and 102 ± 10 g fat per day), balanced by a hospital dietician, in accordance with the Polish standards for hospitalized patients [33]. Detailed nutritional data on the diet during hospitalization, including fiber consumption, are presented in Supplementary Table S1.

This study included 20 patients, with 11 males and 9 females. After admission to the hospital ward, they were all subjected to a 7-day washout from psychiatric medications, received the standard hospital diet, and had a similar hospital routine. The first stool samples were collected after the washout period (W0), and subsequently, OLZ treatment was administered (initially 5 mg/day; doses were individually adjusted up to 20 mg/day). After 6 weeks of treatment, the second stool samples (W6) were collected (Supplementary Figure S1).

Clinical responses were defined as follows: Early responders, 30% reduction in positive and negative syndrome scale (PANNS) total score at 4 weeks; late responders, 40% reduction in PANNS total score at end-point [34]; Clinical global impression-improvement scale (CGI-I) responders, score of 3 points (much improvement); and non-responders, clinical global impression-severity (CGI-S) scores of 4 (minimal improvement) or 5 (no improvement).

2.3. Processing of Raw Data and Statistical Analysis

Sequencing of the V4 region of 16S rRNA gene was performed by the uBiome, Inc. (San Francisco, CA, USA). The 16S amplicons from each sample were individually barcoded and sequenced in the multiplex in the NextSeq 500 platform in a 150 bp (base pair) paired-end modality. The initial quality check of the 16S sequences was conducted using the AfterQC (version 0.9.7) software with default settings [35]. Subsequently, forward and reverse reads were, respectively, capped at 125 and 124 bp and then joined together with an in-between padding sequence (8 of "Ns" with a base score quality of 40). Each sequence was assigned the number of expected errors, and the sequences were filtered to have a maximum expected error of 1.0. The above steps were conducted using the VSEARCH (2.8.0) tool [36].

The sequences were processed using mothur (v.1.41.3) [37]. Briefly, sequences were aligned to the SILVA bacterial reference alignment (release 132), and were then screened to drop those not aligning to positions 13,148 and 25,277 of the SILVA alignment and were pre-clustered to allow two differences between sequences. The chimeras were identified and removed using VSEARCH implemented in mothur. Subsequently, sequences were classified using a Wang method with the Greengenes 16S rRNA Database version 13.8. Finally, sequences were clustered into OTUs using opticlust algorithm and Matthews correlation coefficient metric.

Metagenomic predictions from 16S rRNA marker genes (corrected for predicted 16S rRNA copy number) were carried out using PICRUSt (version 1.1.3) [38], and a list of the KEGG (Kyoto Encyclopedia of Genes and Genomes) functional orthologs (KO) was created. Reference genome coverage of the samples was calculated using the weighted nearest sequenced taxon index (NSTI) [38]. The PICRUSt predicted a median NSTI score of 0.11 (interquartile range, IQR of 0.05). The predicted metagenomes were analyzed with HUMAnN [39] and LEfSe [40]. The KO list was submitted as input data to HUMAnN, which generated KEGG modules and KEGG pathway abundances.

Downstream data analysis was performed using the R software (version 3.5.1, https://cran.r-project.org/), R based tools (such as Phyloseq package (version 1.24.2)) [41] and ComplexHeatmap [42], and custom-made scripts. Before calculating alpha diversity, the samples were rarefied to 3680 sequences per sample. Prior to beta diversity analysis, the taxa with the prevalence of less than 5% were removed (the prevalence of taxa was defined as the proportion of samples in which the taxa appeared at least once). Beta diversity was analyzed using principal coordinate analysis (PCoA) on Bray–Curtis distance matrices generated from the relative OTU abundances. To analyze the changes in bacterial community composition, a change in the principal coordinate 1 (PC1) was examined. The statistical analysis methods included the Wilcoxon rank-sum test, paired Wilcoxon signed-rank test, t-test for one sample, and Spearman rho correlation coefficient. p-values were adjusted using the Benjamini–Hochberg's false discovery rate (FDR) controlling procedure. Numerical data are presented as median, lower quartile, and upper quartile.

3. Results

3.1. Microbiota Compositions

General characteristics of patients are shown in Table 1. There was no significant change in alpha diversity as measured by Chao1 and Shannon indexes ($p = 0.955$ and $p = 0.808$, respectively; Figure 2A). The PCoA with Bray–Curtis dissimilarity is presented in Figure 2B. Samples were separated into distinct regions, mainly along the PC1 (Axis.1) that explained 42.5% of the intersample variance. The gut microbiome was individually specific, and the Bray–Curtis distances between the same samples were significantly smaller than those between all W0 samples ($p = 0.00006$; Figure 2C). The direction of change along the PC1 was not consistent (Supplementary Figure S2). The mean change in the PC1 was not significantly different from 0 (0.0012, (95% confidence interval: −0.0946, 0.0970), $t = −0.03$, df $= 19$, $p = 0.979$), suggesting that the gut microbial community composition does not change after six weeks of treatment. In line with this observation, no OTUs were differentially abundant (from the genus to phylum level) between W0 and W6 (Supplementary Figures S3–S5). There was no change in the ratio of Firmicutes to Bacteroidetes (F/B) in the whole group, as well as in males and females (Supplementary Figure S6). In addition, there were no significant differences in the abundance of the KEGG orthologs, modules, and pathways between W0 and W6 samples in the whole group, as well as in men and women (Supplementary Figure S7).

Table 1. Clinical characteristics of patients included in the study ($n = 20$).

Variables	Median (1st Q–3rd Q)/ n (%)
Sex (F/M)	9 (45%)/11 (55%)
Age (years)	33.5 (31–39)
BMI (kg/m^2)	28.91 (24.82–31.27)
Olanzapine maximum dose (mg)	20.00 (20.0–20.0)
Olanzapine average dose per day (mg)	15.54 (13.50–16.34)
Disease duration (months)	90 (32–114)
Duration of untreated psychosis (months)	4.5 (1.75–12.0)
Smoking (number of cigarettes per day) [a]	1.5 (1.0–3.0)
Coffee (number of cups)	2.0 (0.0–3.0)
Tea (number of cups)	2.5 (1.0–3.0)

BMI—Body Mass Index; [a] Ordinal variables (per day): 1, non-smokers; 2, up to 10 cigarettes; 3, up to 20 cigarettes; 4, up to 40 cigarettes; 1st Q, first quartile; 3rd Q, third quartile, BMI—body mass index.

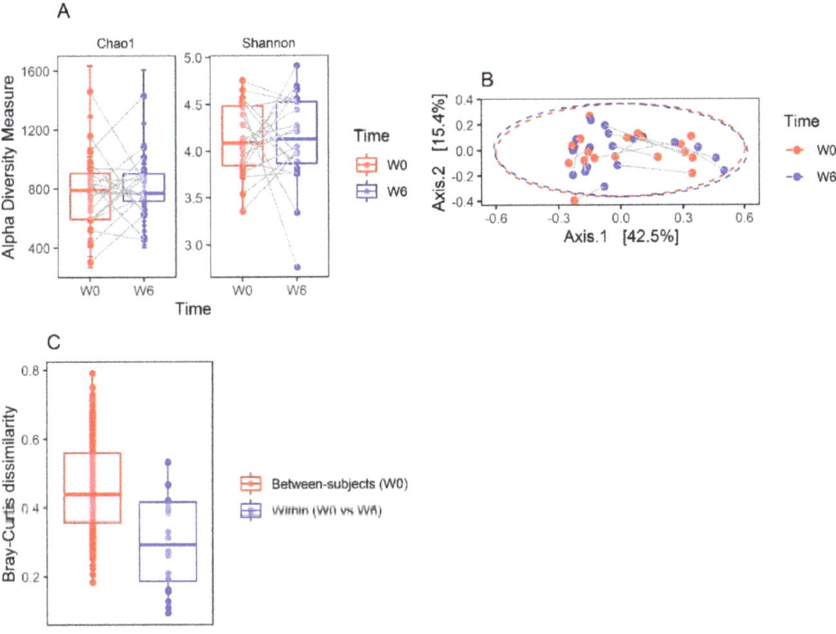

Figure 2. (**A**) Alpha diversity measures at baseline (W0) and after six weeks of hospitalization (W6). The boxplots represent the diversity measures (center line, median; lower and upper hinges correspond to the first (Q1) and third (Q3) quartiles; whiskers, 1.5 * IQR (Q3–Q1). Grey lines connect samples from the same patients. (**B**) Genus level resolution analysis of gut microbiota in patients diagnosed with paranoid schizophrenia treated with olanzapine during six weeks of hospitalization. The principal coordinate analysis was based on Bray–Curtis dissimilarities calculated using relative abundance data. Samples are colored according to time points (W0 and W6). Grey lines connect samples from the same patients. Ellipses correspond to 95% confidence intervals for two timepoints (W0 and W6) with a multivariate normal distribution. (**C**) The boxplot shows Bray–Curtis dissimilarities calculated in the same patients (within (W0 vs. W6), 0.29 (0.19–0.42)) and in different patients (between subjects (W0), 0.44 (0.36–0.56), $p = 0.00006$, Wilcoxon rank-sum test) (center line: median, lower, and upper hinges correspond to the first (Q1) and third (Q3) quartiles; whiskers: the upper whisker is located at the smaller of the maximum Bray–Curtis measures and Q3 + 1.5 * IQR (Q3–Q1); the lower whisker is located at the larger of the minimum Bray–Curtis measures and Q1—1.5 * IQR). W0 and W6 represent time points.

Despite the lack of a consistent shift along the PC1, we examined whether the PC1 changes are associated with demographic, clinical, and environmental factors. The mean PC1 changes did not differ between men (0.011 (−0.118–0.140)) and women (0.004 (−0.036–0.016)) (Wilcoxon rank-sum test FDR adjusted p (q) = 0.649). Demographic, clinical, and environmental factors were not correlated with the change in the PC1 (Supplementary Table S2, Supplementary Figure S8). There was no association between the dose of OLZ and the shift in the gut microbial composition (Supplementary Figure S9). However, the change in the PC1 was significantly greater in patients consuming alcohol (1–3 unit of alcohol; 0.16 (0.03–0.32)) than in non-alcohol drinkers (−0.01 (−0.19–0.01)) (Wilcoxon rank-sum test q = 0.036). To further explore the distinct regions revealed by the ordination of samples by PCoA (Figure 2B), we conducted an unsupervised hierarchical clustering using an average linkage algorithm of the Bray–Curtis dissimilarity. The clustering analysis showed the presence of two clusters (Type 1: 9 samples; Type 2: 31 samples) that matched with distinct regions revealed in the PCoA. A heatmap displaying the relative abundances for the genera annotated with two resulting clusters is shown in Figure 3A. Differential abundance testing revealed that the *Bacteroides*, *Blautia*, *Clostridium*, *Anaerostipes*, *Bilophila*, *Anaerotruncus*, and *Eggerthella* were enriched in the Type 2 cluster, whereas *Prevotella* was enriched in the Type1 cluster (Figure 3B). Clusters Type 1 and 2 seemed to correspond to enterotypes 2 (*Prevotella*) and 1 (*Bacteroides*) described by Arumugam et al. in 2011 [43], respectively. To explore these enterotypes in more detail, our cluster Type 2 was analyzed more thoroughly, as it seemed not to be completely homogenous. Cluster Type 2 was divided into Type 2A and Type 2B, and then the relative abundances of the main contributors from each enterotype in the resulting three clusters (Type 1, Type 2A and 2B) were ascertained (Supplementary Figure S10) Cluster Type 1 had an abundance pattern similar to enterotype 2 (Figure 2d in Arumugam et al. [43]), and that for cluster type 2A to enterotype 1 (Figure 2d in Arumugam et al. [43]). However, cluster type 2B seemed not to be similar to any enterotype. Two genera (*Prevotella* and *Bacteroides*) exhibited similar abundance which was greater than of *Ruminococcus*. The pattern Type 2B seemed to be a type of a mixture from clusters Type 1 and Type 2A. The addition of *Blautia* made no difference to this assessment (Supplementary Figure S11).

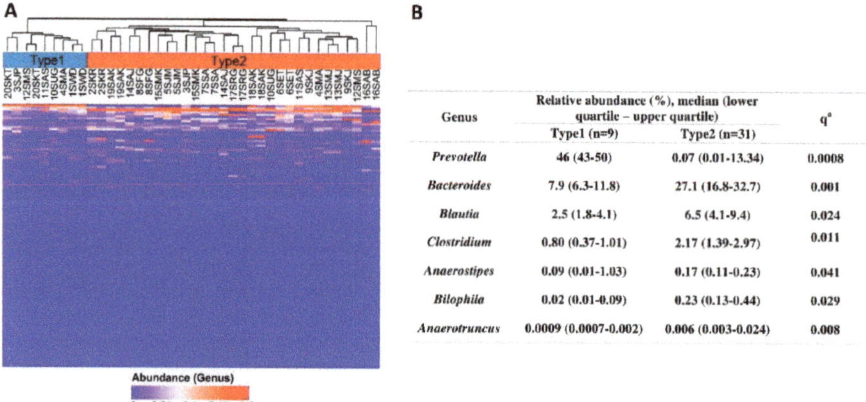

Genus	Relative abundance (%), median (lower quartile – upper quartile)		q^{a}
	Type1 (n=9)	Type2 (n=31)	
Prevotella	46 (43-50)	0.07 (0.01-13.34)	0.0008
Bacteroides	7.9 (6.3-11.8)	27.1 (16.8-32.7)	0.001
Blautia	2.5 (1.8-4.1)	6.5 (4.1-9.4)	0.024
Clostridium	0.80 (0.37-1.01)	2.17 (1.39-2.97)	0.011
Anaerostipes	0.09 (0.01-1.03)	0.17 (0.11-0.23)	0.041
Bilophila	0.02 (0.01-0.09)	0.23 (0.13-0.44)	0.029
Anaerotruncus	0.0009 (0.0007-0.002)	0.006 (0.003-0.024)	0.008

Figure 3. (**A**) Genus level resolution analysis of gut microbiota in patients diagnosed with paranoid schizophrenia treated with olanzapine during six weeks of hospitalization. Unsupervised average linkage hierarchical clustering of gut microbiota at the genus level was conducted. Two resulting clusters (Type 1, blue and Type 2, red) are shown as the top annotation. Both samples (W0 and W6) of 15 patients were found in either Type 1 or Type 2 cluster (two patients in Type 1 and 13 patients in Type 2). Samples of the five patients (3SJP, 4SMA, 10SUG, 11SAS, and 12SMS) belonged to different clusters. (**B**) Differential abundance testing at the genus level between Type 1 and Type 2 clusters. [a] two sided Wilcoxon signed-rank test, FDR adjusted p, the genera with the relative abundance >1% in at least one cluster are shown, *Eggerthella* not shown due to low abundance.

Taken together, our results suggest that the gut microbiota is highly individually specific, and the microbial community compositional changes during six weeks of OLZ treatment are not consistent across the patients.

3.2. Clinical Improvement and BMI Changes

We found that OLZ treatment was associated with significantly improved treatment efficacy as measured by PANNS, 36-item short form survey (SF36), and CGI-S scales (Supplementary Table S3). We further investigated whether these improvements are correlated with the change in microbiota compositions (as measured by a change in the PC1 component) and with demographic and clinical characteristics. No significant correlations were observed between clinical improvements and changes in microbiota composition (Supplementary Figure S12) or demographic and clinical characteristics, except the duration of untreated psychosis (DUP) (Supplementary Table S4).

In contrast to changes in the symptom severity of schizophrenia (Supplementary Table S3), there was no significant change in the patients' BMI during OLZ treatment ($q = 0.763$). However, the BMI change (W6 vs. W0 difference) was significantly higher in women than in men (Supplementary Figure S13) but did not correlate significantly with age, OLZ average dose per day, OLZ maximum dose, disease duration, or duration of untreated psychosis.

Because we found clear differences in gut microbiome compositions in all 40 samples (Figure 3), we next sought to determine whether similar differences in microbial community compositions and metabolic potentials exist in baseline samples and whether those differences could affect the patients' clinical improvement and change in BMI within six weeks. We performed the unsupervised average linkage hierarchical clustering of the Bray–Curtis dissimilarity among the baseline samples (W0, Supplementary Figure S14), as well as that of the relative abundances of the predicted KEGG orthologs, modules, and pathways (Supplementary Figures S15–S17). Regarding the microbiome compositions, we were able to demonstrate different groups of patients (clusters) using hierarchical clustering of KEGG features in the W0 samples: KEGG orthologs (Supplementary Figure S15), modules (Supplementary Figure S16), and pathways (Supplementary Figure S17). Differential abundance testing revealed that only the *Prevotella* genus differed between the two clusters (Type 1, 0.01% (0.006–0.004) vs. Type 2, 27.4% (17.7–43.1); two-sided Wilcoxon signed-rank test, FDR adjusted $p = 0.033$; Supplementary Figure S14). To identify differentially abundant genes, modules, and pathways between clusters, we conducted a linear discriminant analysis with effect size (LEfSe) method (Figure 4).

Subsequently, we compared the baseline symptom scales and BMI between Type 1 and Type 2 clusters. We found significant differences in the baseline PANNS, PANNS G, and CGI-S between the groups created from the clustering of the pathway abundance (Table 2). The patients classified into a Type 2 cluster had significantly more severe symptoms at baseline. The improvement in symptom severity after OLZ treatment assessed by PANNS, SF36, and CG1I was not associated with microbial community compositions (Supplementary Figure S14, Table S5) or KEGG features at baseline (Table 2; Supplementary Figures S15–S17 and Tables S6 and S7). Likewise, no associations were found between baseline gut microbiota (Supplementary Figure S14, Supplementary Table S5) or its metabolic potentials (Table 2 and Supplementary Figures S15–S17 and Supplementary Tables S6 and S7) and the BMI change in the whole group or separately in women or men.

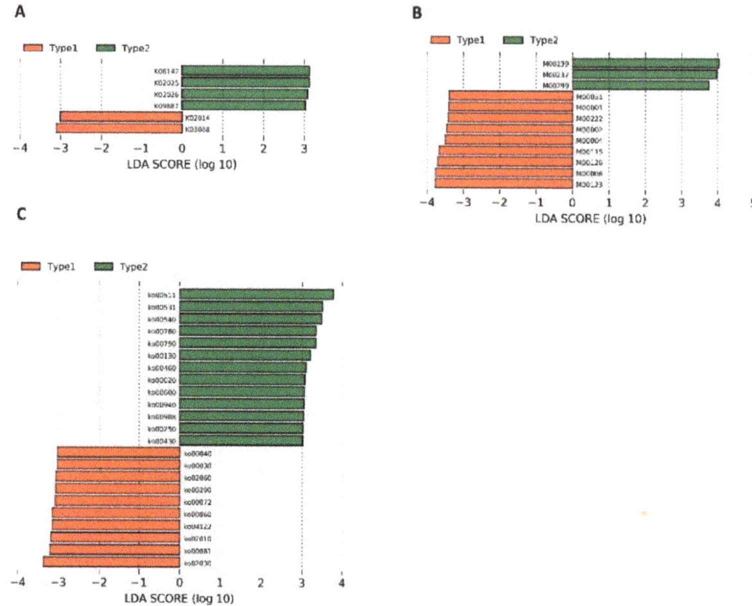

Figure 4. Unsupervised average linkage hierarchical clustering was carried out for each KEGG feature category. (**A**) KEGG orthologs; (**B**) KEGG modules; (**C**) KEGG pathways. K06147, ATP-binding cassette, subfamily B, bacterial; K02025, K02026, multiple sugar transport system permease proteins; K09687, antibiotic transport system ATP-binding protein; K02014, iron complex outer-membrane receptor protein; K03088, RNA polymerase sigma-70 factor, ECF subfamily; M00239, peptides/nickel transport system; M00237, branched-chain amino acid transport system; M00299, Spermidine/putrescine transport system; M00051, Uridine monophosphate biosynthesis, glutamine (+ PRPP) → UMP; M00222, phosphate transport system; M00002, glycolysis, core module involving three-carbon compounds; M00004, pentose phosphate pathway (pentose phosphate cycle); M00115, NAD biosynthesis, aspartate → NAD; M00126, Tetrahydrofolate biosynthesis, GTP → THF; M00006, pentose phosphate pathway, oxidative phase, glucose 6P → ribulose 5P; M00123, Biotin biosynthesis, pimeloyl-CoA → biotin; ko00511, other glycan degradation; ko00531, glycosaminoglycan degradation; ko00540, lipopolysaccharide biosynthesis; ko00780, biotin metabolism; ko00790, folate biosynthesis; ko00130, ubiquinone and other terpenoid–quinone biosynthesis; ko00460, cyanoamino acid metabolism; ko00020, citrate cycle (TCA cycle); ko00600, sphingolipid metabolism; ko00940, phenylpropanoid biosynthesis; ko00908, zeatin biosynthesis; ko00250, alanine, aspartate and glutamate metabolism; ko00430, taurine and hypotaurine metabolism; ko00040, pentose and glucuronate interconversions; ko00030, pentose phosphate pathway; ko02060, phosphotransferase system (PTS); ko00290, valine, leucine and isoleucine biosynthesis; ko00072, synthesis and degradation of ketone bodies; ko00860, porphyrin and chlorophyll metabolism; ko04122, sulfur relay system; ko02010, ABC transporters; ko00061, fatty acid biosynthesis; ko02030, bacterial chemotaxis.

Table 2. Associations of KEGG pathways with BMI changes and clinical improvements (PANNS, SF36, and CGI).

Variables (Females + Males)	Cluster Type 2 ($n = 5$)	Cluster Type 3 ($n = 12$)	p/q[a]
BMI (kg/m^2) W0	28.7 (27–29.9)	29.6 (24.4–32)	0.874/0.874
PANNS W0	95 (94–98)	68 (62.8–74.2)	0.007/0.047
PANNS N subscale W0	28 (23–28)	20.5 (17–22.2)	0.020/0.070
PANNS P subscale W0	24 (23–26)	20 (15.8–22)	0.026/0.073
PANNS G subscale W0	46 (43–47)	32 (27.8–37)	0.010/0.047
SF36 W0	90 (83–97)	76.5 (72.5–83.8)	0.102/0.238
CGI-S W0	7 (6–7)	5 (5–6)	0.009/0.047
BMI (kg/m2)	−0.53 (−1.33–0.72)	0.35 (−0.23–0.90)	0.562/0.656
PANNS	−44 (−65−−31)	−37 (−39.5−−21.8)	0.342/0.749
PANNS N subscale	−10 (−17−−9)	−6 (−8.75−−4.5)	0.205/0.410
PANNS P subscale	−12 (−18−−6)	−11 (−15−−7.75)	0.874/0.874
PANNS G subscale	−22 (−26−−16)	−16 (−17.8−−8.75)	0.315/0.479
SF36	−5 (−18−−4)	−3 (−12.2–6)	0.245/0.429
CGI-I	4 (3–4)	3.5 (3–4)	0.452/0.575
Variable (Males)	**Cluster Type 2 ($n = 5$)**	**Cluster Type 3 ($n = 4$)**	**p/q[a]**
BMI (kg/m^2) W0	28.7 (27–29.9)	30.5 (27.5–32.2)	0.713/0.768
PANNS W0	95 (94–98)	67 (59.2–76.5)	0.037/0.198
PANNS N subscale W0	28 (23–28)	20 (18.8–22.2)	0.084/0.198
PANNS P subscale W0	24 (23–26)	17 (14–19)	0.027/0.198
PANNS G subscale W0	46 (43–47)	30.5 (24.8–37.5)	0.065/0.198
SF36 W0	90 (83–97)	83.5 (76.2–90)	0.391/0.547
CGI-S W0	7 (6–7)	5.5 (5–6)	0.050/0.198
BMI (kg/m^2)	−0.53 (−1.33–0.72)	−0.92 (−1.97−−0.30)	0.713/0.768
PANNS	−44 (−65−−31)	−20.5 (−26.2−−17.8)	0.140/0.280
PANNS N subscale	−10 (−17−−9)	−6 (−7.5−−5.75)	0.389/0.547
PANNS P subscale	−12 (−18−−6)	−7.5 (−8.25−−6)	0.389/0.547
PANNS G subscale	−22 (−26−−16)	−8.5 (−11−−7)	0.085/0.198
SF36	−5 (−18−−4)	−11 (−13.5−−7)	1.0/1.0
CGI-I	4 (3–4)	4 (4–4)	0.661/0.768

[a] Two-sided Wilcoxon rank-sum test, median with lower and upper quartiles in parentheses; BMI, PANNS, and SF36—changes from baseline (W0); CGI-I—an improvement from baseline; KEGG, Kyoto Encyclopedia of Genes and Genomes; BMI, body mass index; PANNS, positive and negative syndrome scale; SF36, 36- item short form survey; CGI, clinical global impression-improvement scale.

To further explore the gut microbiota and OLZ treatment interactions, we classified the included patients as responders and non-responders as follows: Early responders, early non-responders, late responders, and late non-responders using the PANNS total score and responders and non-responders using the CGI-I scale. Subsequently, microbial community compositions and KEGG features were compared between responders and non-responders. Phylogenetic compositions of the samples at the phylum level in the responders and non-responders are shown in Supplementary Figure S18. The phyla were not differentially abundant in responders and non-responders, regardless of the definition of clinical improvement. There were no differences in gut microbiome compositions at other taxonomic levels (Supplementary Figure S19), as well as in the KEGG orthologs, modules, and pathways (Supplementary Figure S20). Sex-specific results are shown in Supplementary Figures S21 and S22 (bacterial community composition) and S23 and S24 (KEGG features).

4. Discussion

The effect of OLZ on the microbiota has been investigated in experimental studies. Davey et al. [29,30] found decreased gut microbiota diversity, increased abundance of phyla Firmicutes, and reduced Actinobacteria, Proteobacteria, and Bacteroidetes in the course of OLZ treatment in female rats. Similarly, Morgan et al. [31] revealed decreased alpha diversity, lower abundance of class

Bacteroidia, and increased abundances of Erysipelotrichia, Actinobacteria, and Gammaproteobacteria in female mice treated with OLZ. However, Kao et al. [44] demonstrated no significant effects of OLZ on gut microbiota in female rats. To the best of our knowledge, this study is the first to analyze fecal microbiota compositions in patients hospitalized due to acute relapse of SZ. We did not find the impact of six-week OLZ treatment on bacterial diversity, abundance, and predicted metabolic function, and patients with SZ had individualized and stable gut microbiota in the course of six-week OLZ treatment in terms of both composition and function. Because of the inconsistent findings above, further studies are needed to clarify the effect of OLZ on gut microbiota.

Although gut microbiota could be compositionally and functionally clustered into similar groups, the classification could not be used to predict the responses to OLZ treatment or the occurrence of weight gain (observed only in women) during OLZ treatment. As mentioned above, OLZ causes weight gain in female rats [29,44] and mice [31]. This metabolic effect is not observed during antibiotic therapy [29] and gnotobiosis (germ-free mouse model) and is enhanced during the administration of the high-fat diet regimen that is responsible for alterations of microbiota similar to those observed in metabolic syndromes [31]. In addition, Davey et al. [30] demonstrated metabolic disturbances, inflammation, and microbiota alterations in female mice treated with OLZ and found only slight alterations in male mice treated with OLZ, and metabolic effects of OLZ were linked to gut microbiota alterations. Notably, antibiotics reversed these effects due to reduced gut microbiota. Therefore, changed gut microbiota plays a pivotal role in weight gain. The lack of association between fecal microbiota compositions and weight gain in this study may be due to the low number of participants and the short period of OLZ administration. In addition, other factors might also be responsible for the increase in body mass index associated with the OLZ administration [20,26–28].

In the present study two bacterial enterotypes (clusters) were found, Type 1, with a predominance of *Prevotella*, and Type 2 with a higher abundance of *Bacteroides*, *Blautia* and *Clostridium*. Cluster Type 2 seemed not to be completely homogenous (with Types 2A and 2B), which initially suggested the possibility of the occurrence of a third enterotype similar to that found by Arumugam et al. [43]. Further analysis did not confirm this hypothesis and a higher abundance of *Ruminococcus* or *Blautia* in sub-cluster Type 2B was not seen. This sub-cluster seemed to be a type of mixture from clusters Type 1 and Type 2A. Due to this we took into consideration in further analyses only two enterotypes (original clusters) of bacteria. Moreover, patients with SZ were clustered at the level of KEGG genes, modules, and pathways. The severity of symptoms measured at the beginning of treatment varied, depending on the predicted metabolic activity of the bacteria. Other studies also have observed a relationship between the composition of bacteria and the severity of symptoms in SZ patients. Zheng et al. [11] demonstrated that PANSS was negatively correlated with Veillonellaceae and was positively correlated with Bacteroidaceae, Streptococcaceae, and Lachnospiraceae. Schwartz et al. [10] found greater microbial abnormalities in SZ patients than in controls. In addition, increases in the number of Lactobacillus group bacteria were positively correlated with the severity of various symptom domains in SZ patients and were negatively correlated with the global assessment of functioning. Moreover, responses to the treatment were worse in patients with severe microbiota alterations. Furthermore, Shen et al. [45], using the PICRUSt analysis, infer that vitamin B6 and fatty acid metabolic potential differed significantly between SZ patients and controls. Therefore, there are potential relationships between predicted metabolic changes and the severity of symptoms in SZ patients, as shown in Table 3. It is important to note that the PICRUSt approach using in prediction of bacterial metabolic activity should be treated with caution and followed by metagenomic analyses to explain such findings in humans. The median NSTI score was 0.11 (interquartile range of 0.05) suggesting a reasonable accuracy of the prediction, however, some closely related reference genomes were not available.

Table 3. Potential relationships between predicted metabolic changes and the severity of symptoms in schizophrenia (SZ) patients.

Pathways	Physiological Function	Potential Roles in SZ	References
Pathways found to be more active in patients with significantly less severe symptoms (according to PANNS and CGI-S)			
ko00430: Taurine and hypotaurine metabolism	Taurine: N-methyl-D-aspartate (NDMA) receptor inhibition and stem cell activation; a neurotransmitter and an inhibitory neuromodulator in the central nervous system (CNS); a potential immunomodulating compound, and an attenuator of apoptosis	Taurine supplementation was found to alleviate SZ symptoms significantly	[46–48]
Ko00250: Alanine (ALA), aspartate (ASP), and glutamate metabolism	ALA: An agonist that binds to the glycine site of NMDA receptors and improves the positive and cognitive symptoms of patients with SZ; ASP: Binding to the agonist site of NMDARs	NMDAR hypofunction in schizophrenia pathogenesis	[49,50]
Ko00790: Folate biosynthesis	Folate: Production of adenosylmethionine (SAM)	Schizophrenia patients may have lower folate levels (negative correlation with negative symptoms of SZ)	[51–53]
Ko00130: Ubiquinone and other terpenoids–quinone biosynthesis	Ubiquinone: ATP production, mitochondrial function, and reduction of proinflammatory mediators	Mitochondrial dysfunction as a part of SZ etiology	[54,55]
Ko00020: Citrate cycle (TCA cycle)	TCA: Normal energy metabolism of the brain	Abnormalities in energy metabolism were found to play a role in SZ pathophysiology	[56]
Ko00600: Sphingolipid (SL) metabolism	Formation of membrane "lipid rafts" of myelin sheaths, especially in neurons and oligodendrocytes (crucial for normal synaptic neurotransmission, axon-myelin stability, and communication/connectivity)	Inflammatory, synaptic, and white matter changes that result in disconnectivity in SZ may be related to SL	[57,58]
Pathways found to be more active in patients with significantly more severe symptoms (according to PANNS and CGI-S)			
Ko00030: Pentose phosphate pathway	Formation of NADPH for biosynthetic processes, cellular redox balance, and synthesis of ribose	Pentose phosphate pathway-related molecules in schizophrenia were found to be increased	[59]
Ko00061: Fatty acid biosynthesis	Component of membranes and myelination process mediator	Lipolysis and β-oxidation were found to be upregulated in SZ, as a result of insufficient brain energy supply	[60,61]
Ko00290: Valine, leucine, and isoleucine biosynthesis	Protein synthesis, production of energy, compartmentalization of glutamate synthesis of amine neurotransmitters, including serotonin, dopamine, and norepinephrine	Branched- chain amino acids when administered to patients with tardive dyskinesia—aberration of voluntary motor control in SZ patients treated with psychotropic drugs	[62,63]
Ko00072: Synthesis and degradation of ketone bodies	An alternative source of energy under fasting and starving; restrictive diets prolonged intense exercise	Ketones may change the ratio of GABA (glutamate in favor of GABA) to compensate GABA levels in the CNS in SZ patients	[64]

PANNS—The Positive and Negative Syndrome Scale, CGI-S—The Clinical Global Impressions Scale.

Our study has several strengths that should be highlighted. (1) The applied treatment resulted in expected clinical effects. The relationship between duration of untreated psychosis (DUP) and poor general symptomatic outcomes was confirmed, and the longer DUP was associated with more severe positive and negative symptoms. Additionally, OLZ treatment caused weight gain. This observation is in agreement with that in another experimental study [65]. (2) During the treatment, the patients were under the same controlled hospital conditions (diet, drug intake, and clinical monitoring), and a washout was used before treatment, thus providing a "unification" of the environmental impact on the fecal microbiota pattern. Consequently, we speculate that such conditions diminish the impact of common environmental factors that permanently shape gut microbiota composition and underline the association between the disease and treatment. (3) Weight gain at the beginning of OLZ treatment is very important because it determines the further development of cardiometabolic risk factors [22,23,66]. (4) Although the study group was not homogeneous, the symptoms were observed every day in our psychiatric clinic. Patients were previously treated with other pharmaceuticals, which might have affected the microbiota composition. Such situations might lead to resistance against psychotropic drugs, probably leading to no impact of OLZ on the microbiota.

There are certain limitations of our study that need to be discussed. First, the sample size was small and heterogeneous (drug-naive and previously-treated patients). No formal sample size calculations were employed for this analysis, but the cohort size was based on what was previously sufficient to test microbiotic changes in schizophrenia patients [67], and/or the influence of antibiotics [68] and risperidone administration [69] on gut microbiota. This limitation should be attributed to rigorous inclusion and exclusion criteria as well as the short duration of the whole study (17 months). Thus, studies with a greater sample size are needed to further examine the associations between OLZ treatment and gut microbiota structure. Second, the composition of intestinal bacteria varied among individuals, and inter-individual variation within the gut ecosystem of patients was high. Third, in individual studies (also experimental), various taxonomic groups of bacteria were analyzed only in stools. The composition of bacteria in feces is more stable and is not influenced by external factors compared with the composition of bacteria in the small intestine. Changes in the microbiota of the small bowel have a much greater effect on the metabolic functions of the human body. Therefore, further experimental studies should pay more attention to this issue [70,71], although an invasive way of sampling intestine biological material remains difficult and holds several ethical concerns. Fourth, there was a lack of long-term follow-up, which is especially important in case of metabolic consequences of OLZ treatment. Fifth, we did not compare the results between SZ patients and healthy subjects or patients receiving placebos. Matched controls with similar lifestyle should be used to exclude false-positive results. However, the general lifestyle in patients diagnosed with SZ was found to be divergent from that observed in healthy people [72]. Therefore, a placebo approach was impossible mainly due to ethical and organizational concerns. Sixth, changes in dietary and living conditions during the hospital stay might be another limitation of our study. However, enterotypes *Prevotella* and *Bacteroides* are strongly associated with long-term diet. It was shown that microbiome composition changed detectably within 24 h of initiating a high-fat/low-fiber or low-fat/high-fiber diet, but that enterotype identity remained stable during the 10-day study [73]. Therefore, a change of diet after admission to hospital should not affect W0 microbiota. After this all patients received the same diet, and it seems that this factor should also not significantly affect the influence of OLZ on W6 microbiota composition.

5. Conclusions

In conclusion, the present findings indicate that the microbiota in patients with the schizophrenia episode is highly individualized, although it can be clustered into different taxonomical (Type 1, with a predominance of *Prevotella*, and Type 2 with a higher abundance of *Bacteroides*, *Blautia*, and *Clostridium*) and functional groups; the microbiota does not change during six weeks of treatment with OLZ and is not associated with the weight gain that occurs in women treated with OLZ, as well as the treatment effectiveness. This study provides some insights into the metabolic effects of psychotropic drugs on gut microbiota in SZ patients. Further long-term and placebo-controlled studies are needed to clarify the effect of OLZ on gut microbiota.

Supplementary Materials: The following are available online at http://www.mdpi.com/2077-0383/8/10/1605/s1.

Author Contributions: Conceptualization, J.P.-W., A.B.-K. and J.S.; data curation, J.P.-W., M.K., A.B.-K., P.L., M.W. and I.Ł.; formal analysis, J.P.-W., M.K., P.L., K.S.-Ż., W.M., B.M., J.K.-M., I.Ł. and J.S.; investigation, J.P.-W., M.K., A.B.-K., P.L., M.W., K.S.-Ż., W.M., J.K.-M., I.Ł. and J.S.; methodology, J.S.; software, M.K.; supervision, J.P.-W. and J.S.; visualization, M.K.; writing—original draft, J.P.-W., M.K. and I.Ł.; writing—review and editing, J.P.-W., M.K., A.B.-K., M.W., K.S.-Ż., W.M., B.M., T.S., I.Ł. and J.S.

Funding: The study was funded within the framework of the project Fundusz Stymulacji Nauki (grant number: FSN-312-05/15) provided by Pomeranian Medical University at Szczecin (Poland) and National Science Centre (UMO-2018/31/B/NZ5/00527). The funding sources had no role in concept design, selection of articles, the decision to publish, or the preparation of the manuscript.

Acknowledgments: We would like to thank Editage (www.editage.com) for English language editing.

Conflicts of Interest: I.Ł. and W.M. are cofounders and shareholders at Sanprobi company. However, the content of this study was not constrained by this fact. Other authors declare no conflicts of interest.

References

1. WHO Schizophrenia. Available online: https://www.who.int/news-room/fact-sheets/detail/schizophrenia (accessed on 2 June 2019).
2. Müller, N. Inflammation in Schizophrenia: Pathogenetic Aspects and Therapeutic Considerations. *Schizophr. Bull.* **2018**, *44*, 973–982. [CrossRef] [PubMed]
3. Purves-Tyson, T.D.; Owens, S.J.; Rothmond, D.A.; Halliday, G.M.; Double, K.L.; Stevens, J.; McCrossin, T.; Weickert, C.S. Putative presynaptic dopamine dysregulation in schizophrenia is supported by molecular evidence from post-mortem human midbrain. *Transl. Psychiatry* **2017**, *7*, e1003. [CrossRef] [PubMed]
4. Guest, F.L.; Martins-de-Souza, D.; Rahmoune, H.; Bahn, S.; Guest, P.C. The effects of stress on hypothalamic-pituitary-adrenal (HPA) axis function in subjects with schizophrenia. *Arch. Clin. Psychiatry (São Paulo)* **2013**, *40*, 20–27. [CrossRef]
5. Guest, P.C.; Martins-De-Souza, D.; Vanattou-Saifoudine, N.; Harris, L.W.; Bahn, S. Abnormalities in Metabolism and Hypothalamic–Pituitary–Adrenal Axis Function in Schizophrenia. *Int. Rev. Neurobiol.* **2011**, *101*, 145–168.
6. Clemente, J.C.; Manasson, J.; Scher, J.U. The role of the gut microbiome in systemic inflammatory disease. *BMJ* **2018**, *360*, j5145. [CrossRef]
7. Sudo, N.; Chida, Y.; Aiba, Y.; Sonoda, J.; Oyama, N.; Yu, X.-N.; Kubo, C.; Koga, Y. Postnatal microbial colonization programs the hypothalamic–pituitary–adrenal system for stress response in mice. *J. Physiol.* **2004**, *558*, 263–275. [CrossRef]
8. Waclawiková, B.; El Aidy, S. Role of Microbiota and Tryptophan Metabolites in the Remote Effect of Intestinal Inflammation on Brain and Depression. *Pharmaceuticals (Basel)* **2018**, *11*, 63. [CrossRef]
9. Rogers, G.B.; Keating, D.J.; Young, R.L.; Wong, M.-L.; Licinio, J.; Wesselingh, S. From gut dysbiosis to altered brain function and mental illness: Mechanisms and pathways. *Mol. Psychiatry* **2016**, *21*, 738–748. [CrossRef]
10. Schwarz, E.; Maukonen, J.; Hyytiäinen, T.; Kieseppä, T.; Orešič, M.; Sabunciyan, S.; Mantere, O.; Saarela, M.; Yolken, R.; Suvisaari, J. Analysis of microbiota in first episode psychosis identifies preliminary associations with symptom severity and treatment response. *Schizophr. Res.* **2018**, *192*, 398–403. [CrossRef]
11. Zheng, P.; Zeng, B.; Liu, M.; Chen, J.; Pan, J.; Han, Y.; Liu, Y.; Cheng, K.; Zhou, C.; Wang, H.; et al. The gut microbiome from patients with schizophrenia modulates the glutamate-glutamine-GABA cycle and schizophrenia-relevant behaviors in mice. *Sci. Adv.* **2019**, *5*, eaau8317. [CrossRef]
12. Rakoff-Nahoum, S.; Paglino, J.; Eslami-Varzaneh, F.; Edberg, S.; Medzhitov, R. Recognition of Commensal Microflora by Toll-Like Receptors Is Required for Intestinal Homeostasis. *Cell* **2004**, *118*, 229–241. [CrossRef] [PubMed]
13. Barry, S.; Clarke, G.; Scully, P.; Dinan, T.G. Kynurenine pathway in psychosis: Evidence of increased tryptophan degradation. *J. Psychopharmacol.* **2009**, *23*, 287–294. [CrossRef] [PubMed]

14. Miller, C.L.; Llenos, I.C.; Dulay, J.R.; Weis, S. Upregulation of the initiating step of the kynurenine pathway in postmortem anterior cingulate cortex from individuals with schizophrenia and bipolar disorder. *Brain Res.* **2006**, *1073*, 25–37. [CrossRef] [PubMed]
15. Hálfdánarson, Ó.; Zoëga, H.; Aagaard, L.; Bernardo, M.; Brandt, L.; Fusté, A.C.; Furu, K.; Garuoliené, K.; Hoffmann, F.; Huybrechts, K.F.; et al. International trends in antipsychotic use: A study in 16 countries, 2005–2014. *Eur. Neuropsychopharmacol.* **2017**, *27*, 1064–1076.
16. Vancampfort, D.; Correll, C.U.; Galling, B.; Probst, M.; De Hert, M.; Ward, P.B.; Rosenbaum, S.; Gaughran, F.; Lally, J.; Stubbs, B. Diabetes mellitus in people with schizophrenia, bipolar disorder and major depressive disorder: A systematic review and large scale meta-analysis. *World Psychiatry* **2016**, *15*, 166–174. [CrossRef]
17. Galling, B.; Roldán, A.; Nielsen, R.E.; Nielsen, J.; Gerhard, T.; Carbon, M.; Stubbs, B.; Vancampfort, D.; De Hert, M.; Olfson, M.; et al. Type 2 Diabetes Mellitus in Youth Exposed to Antipsychotics. *JAMA Psychiatry* **2016**, *73*, 247–259. [CrossRef]
18. Galling, B.; Correll, C.U. Do antipsychotics increase diabetes risk in children and adolescents? *Expert Opin. Drug Saf.* **2015**, *14*, 219–241. [CrossRef]
19. De Hert, M.; Vancampfort, D.; Correll, C.U.; Mercken, V.; Peuskens, J.; Sweers, K.; Van Winkel, R.; Mitchell, A.J. Guidelines for screening and monitoring of cardiometabolic risk in schizophrenia: Systematic evaluation. *Br. J. Psychiatry* **2011**, *199*, 99–105. [CrossRef]
20. Bak, M.; Fransen, A.; Janssen, J.; Van Os, J.; Drukker, M. Almost All Antipsychotics Result in Weight Gain: A Meta-Analysis. *PLoS ONE* **2014**, *9*, e94112. [CrossRef]
21. Musil, R.; Obermeier, M.; Russ, P.; Hamerle, M. Weight gain and antipsychotics: A drug safety review. *Expert Opin. Drug Saf.* **2015**, *14*, 73–96. [CrossRef]
22. Spertus, J.; Horvitz-Lennon, M.; Abing, H.; Normand, S.-L. Risk of weight gain for specific antipsychotic drugs: A meta-analysis. *NPJ Schizophr.* **2018**, *4*, 12. [CrossRef] [PubMed]
23. Pérez-Iglesias, R.; Martínez-García, O.; Pardo-Garcia, G.; Amado, J.A.; Garcia-Unzueta, M.T.; Tabares-Seisdedos, R.; Crespo-Facorro, B. Course of weight gain and metabolic abnormalities in first treated episode of psychosis: The first year is a critical period for development of cardiovascular risk factors. *Int. J. Neuropsychopharmacol.* **2014**, *17*, 41–51. [CrossRef] [PubMed]
24. American Diabetes Association; American Psychiatric Association; American Association of Clinical Endocrinologists; North American Association for the Study of Obesity Consensus development conference on antipsychotic drugs and obesity and diabetes. *Diabetes Care* **2004**, *27*, 596–601.
25. De Hert, M.; Dekker, J.M.; Wood, D.; Kahl, K.G.; Holt, R.I.G.; Möller, H.-J. Cardiovascular disease and diabetes in people with severe mental illness position statement from the European Psychiatric Association (EPA), supported by the European Association for the Study of Diabetes (EASD) and the European Society of Cardiology (ESC). *Eur. Psychiatry* **2009**, *24*, 412–424. [CrossRef]
26. Albaugh, V.L.; Henry, C.R.; Bello, N.T.; Hajnal, A.; Lynch, S.L.; Halle, B.; Lynch, C.J. Hormonal and metabolic effects of olanzapine and clozapine related to body weight in rodents. *Obesity (Silver Spring)* **2006**, *14*, 36–51. [CrossRef]
27. Basson, B.R.; Kinon, B.J.; Taylor, C.C.; Szymanski, K.A.; Gilmore, J.A.; Tollefson, G.D. Factors Influencing Acute Weight Change in Patients with Schizophrenia Treated with Olanzapine, Haloperidol, or Risperidone. *J. Clin. Psychiatry* **2001**, *62*, 231–238. [CrossRef]
28. Ou, J.-J.; Xu, Y.; Chen, H.-H.; Fan, X.; Gao, K.; Wang, J.; Guo, X.-F.; Wu, R.-R.; Zhao, J.-P. Comparison of metabolic effects of ziprasidone versus olanzapine treatment in patients with first-episode schizophrenia. *Psychopharmacology* **2013**, *225*, 627–635. [CrossRef]
29. Davey, K.J.; Cotter, P.D.; O'Sullivan, O.; Crispie, F.; Dinan, T.G.; Cryan, J.F.; O'Mahony, S.M. Antipsychotics and the gut microbiome: Olanzapine-induced metabolic dysfunction is attenuated by antibiotic administration in the rat. *Transl. Psychiatry* **2013**, *3*, e309. [CrossRef]
30. Davey, K.J.; O'Mahony, S.M.; Schellekens, H.; O'Sullivan, O.; Bienenstock, J.; Cotter, P.D.; Dinan, T.G.; Cryan, J.F. Gender-dependent consequences of chronic olanzapine in the rat: Effects on body weight, inflammatory, metabolic and microbiota parameters. *Psychopharmacology* **2012**, *221*, 155–169. [CrossRef]
31. Morgan, A.P.; Crowley, J.J.; Nonneman, R.J.; Quackenbush, C.R.; Miller, C.N.; Ryan, A.K.; Bogue, M.A.; Paredes, S.H.; Yourstone, S.; Carroll, I.M.; et al. The Antipsychotic Olanzapine Interacts with the Gut Microbiome to Cause Weight Gain in Mouse. *PLoS ONE* **2014**, *9*, e115225. [CrossRef]

32. Flowers, S.A.; Evans, S.J.; Ward, K.M.; McInnis, M.G.; Ellingrod, V.L. Interaction Between Atypical Antipsychotics and the Gut Microbiome in a Bipolar Disease Cohort. *Pharmacother. J. Hum. Pharmacol. Drug Ther.* **2017**, *37*, 261–267. [CrossRef] [PubMed]

33. Jarosz, M.; Rychlik, E.; Stoś, K.; Wierzejska, R.; Wojtasik, A.; Charzewska, J.; Mojska, H.; Szponar, L.; Sajór, I.; Kłosiewicz-Latoszek, L.; et al. *Normy Żywienia dla Populacji Polski*; Instytut Żywności i Żywienia: Warszawa, Poland, 2017; ISBN 978-83-86060-89-4.

34. Ascher-Svanum, H.; Zhao, F.; Detke, H.C.; Nyhuis, A.W.; Lawson, A.H.; Stauffer, V.L.; Montgomery, W.; Witte, M.M.; McDonnell, D.P. Early response predicts subsequent response to olanzapine long-acting injection in a randomized, double-blind clinical trial of treatment for schizophrenia. *BMC Psychiatry* **2011**, *11*, 152. [CrossRef] [PubMed]

35. Chen, S.; Huang, T.; Zhou, Y.; Han, Y.; Xu, M.; Gu, J. AfterQC: Automatic filtering, trimming, error removing and quality control for fastq data. *BMC Bioinform.* **2017**, *18*, 80. [CrossRef] [PubMed]

36. Rognes, T.; Flouri, T.; Nichols, B.; Quince, C.; Mahé, F. VSEARCH: A versatile open source tool for metagenomics. *PeerJ* **2016**, *4*, e2584. [CrossRef] [PubMed]

37. Schloss, P.D.; Westcott, S.L.; Ryabin, T.; Hall, J.R.; Hartmann, M.; Hollister, E.B.; Lesniewski, R.A.; Oakley, B.B.; Parks, D.H.; Robinson, C.J.; et al. Introducing mothur: Open-Source, Platform-Independent, Community-Supported Software for Describing and Comparing Microbial Communities. *Appl. Environ. Microbiol.* **2009**, *75*, 7537–7541. [CrossRef]

38. Langille, M.G.I.; Zaneveld, J.; Caporaso, J.G.; McDonald, D.; Knights, D.; Reyes, J.A.; Clemente, J.C.; Burkepile, D.E.; Thurber, R.L.V.; Knight, R.; et al. Predictive functional profiling of microbial communities using 16S rRNA marker gene sequences. *Nat. Biotechnol.* **2013**, *31*, 814–821. [CrossRef]

39. Abubucker, S.; Segata, N.; Goll, J.; Schubert, A.M.; Izard, J.; Cantarel, B.L.; Rodriguez-Mueller, B.; Zucker, J.; Thiagarajan, M.; Henrissat, B.; et al. Metabolic Reconstruction for Metagenomic Data and Its Application to the Human Microbiome. *PLoS Comput. Biol.* **2012**, *8*, e1002358. [CrossRef]

40. Segata, N.; Izard, J.; Waldron, L.; Gevers, D.; Miropolsky, L.; Garrett, W.S.; Huttenhower, C. Metagenomic biomarker discovery and explanation. *Genome Biol.* **2011**, *12*, R60. [CrossRef]

41. McMurdie, P.J.; Holmes, S. phyloseq: An R Package for Reproducible Interactive Analysis and Graphics of Microbiome Census Data. *PLoS ONE* **2013**, *8*, e61217. [CrossRef]

42. Gu, Z.; Eils, R.; Schlesner, M. Complex heatmaps reveal patterns and correlations in multidimensional genomic data. *Bioinformatics* **2016**, *32*, 2847–2849. [CrossRef]

43. Arumugam, M.; Raes, J.; Pelletier, E.; Le Paslier, D.; Yamada, T.; Mende, D.R.; Fernandes, G.R.; Tap, J.; Bruls, T.; Batto, J.-M.; et al. Enterotypes of the human gut microbiome. *Nature* **2011**, *473*, 174–180. [CrossRef] [PubMed]

44. Kao, A.C.-C.; Spitzer, S.; Anthony, D.C.; Lennox, B.; Burnet, P.W.J. Prebiotic attenuation of olanzapine-induced weight gain in rats: Analysis of central and peripheral biomarkers and gut microbiota. *Transl. Psychiatry* **2018**, *8*, 66. [CrossRef] [PubMed]

45. Shen, Y.; Xu, J.; Li, Z.; Huang, Y.; Yuan, Y.; Wang, J.; Zhang, M.; Hu, S.; Liang, Y. Analysis of gut microbiota diversity and auxiliary diagnosis as a biomarker in patients with schizophrenia: A cross-sectional study. *Schizophr. Res.* **2018**, *197*, 470–477. [CrossRef] [PubMed]

46. Firth, J.; Rosenbaum, S.; Ward, P.B.; Curtis, J.; Teasdale, S.B.; Yung, A.R.; Sarris, J. Adjunctive nutrients in first-episode psychosis: A systematic review of efficacy, tolerability and neurobiological mechanisms. *Early Interv. Psychiatry* **2018**, *12*, 774–783. [CrossRef] [PubMed]

47. Leppik, L.; Kriisa, K.; Koido, K.; Koch, K.; Kajalaid, K.; Haring, L.; Vasar, E.; Zilmer, M. Profiling of Amino Acids and Their Derivatives Biogenic Amines Before and After Antipsychotic Treatment in First-Episode Psychosis. *Front. Psychol.* **2018**, *9*, 155. [CrossRef] [PubMed]

48. O'Donnell, C.P.; Allott, K.A.; Murphy, B.P.; Yuen, H.P.; Proffitt, T.-M.; Papas, A.; Moral, J.; Pham, T.; O'Regan, M.K.; Phassouliotis, C.; et al. Adjunctive Taurine in First-Episode Psychosis: A Phase 2, Double-Blind, Randomized, Placebo-Controlled Study. *J. Clin. Psychiatry* **2016**, *77*, e1610–e1617. [CrossRef] [PubMed]

49. Hatano, T.; Ohnuma, T.; Sakai, Y.; Shibata, N.; Maeshima, H.; Hanzawa, R.; Suzuki, T.; Arai, H. Plasma alanine levels increase in patients with schizophrenia as their clinical symptoms improve—Results from the Juntendo University Schizophrenia Projects (JUSP). *Psychiatry Res. Neuroimaging* **2010**, *177*, 27–31. [CrossRef]

50. Errico, F.; Nuzzo, T.; Carella, M.; Bertolino, A.; Usiello, A. The Emerging Role of Altered d-Aspartate Metabolism in Schizophrenia: New Insights from Preclinical Models and Human Studies. *Front. Psychol.* **2018**, *9*, 559. [CrossRef]

51. Cao, B.; Wang, D.-F.; Xu, M.-Y.; Liu, Y.-Q.; Yan, L.-L.; Wang, J.-Y.; Lu, Q.-B. Lower folate levels in schizophrenia: A meta-analysis. *Psychiatry Res.* **2016**, *245*, 1–7. [CrossRef]

52. Hussein, H.; El Mawella, S.A.; Ahmed, T. Folate, vitamin B12, and negative symptoms in schizophrenia. *Egypt. J. Psychiatry* **2018**, *39*, 89. [CrossRef]

53. Hill, M.; Shannahan, K.; Jasinski, S.; Macklin, E.A.; Raeke, L.; Roffman, J.L.; Goff, D.C. Folate supplementation in schizophrenia: A possible role for MTHFR genotype. *Schizophr. Res.* **2011**, *127*, 41–45. [CrossRef] [PubMed]

54. Maguire, Á.; Hargreaves, A.; Gill, M. Coenzyme Q10 and neuropsychiatric and neurological disorders: Relevance for schizophrenia. *Nutr. Neurosci.* **2018**, *21*, 1–14. [CrossRef] [PubMed]

55. Schmelzer, C.; Döring, F. Identification of LPS-inducible genes downregulated by ubiquinone in human THP-1 monocytes. *BioFactors* **2010**, *36*, 222–228. [CrossRef] [PubMed]

56. Bubber, P.; Hartounian, V.; Gibson, G.E.; Blass, J.P. Abnormalities in the tricarboxylic acid (TCA) cycle in the brains of schizophrenia patients. *Eur. Neuropsychopharmacol.* **2011**, *21*, 254–260. [CrossRef]

57. Castillo, R.I.; Rojo, L.E.; Henríquez-Henríquez, M.; Silva, H.; Maturana, A.; Villar, M.J.; Fuentes, M.; Gaspar, P.A. From Molecules to the Clinic: Linking Schizophrenia and Metabolic Syndrome through Sphingolipids Metabolism. *Front. Mol. Neurosci.* **2016**, *10*, 488. [CrossRef]

58. Weston-Green, K.; Babic, I.; De Santis, M.; Pan, B.; Montgomery, M.K.; Mitchell, T.; Huang, X.-F.; Nealon, J. Disrupted sphingolipid metabolism following acute clozapine and olanzapine administration. *J. Biomed. Sci.* **2018**, *25*, 40. [CrossRef]

59. Liu, M.-L.; Zhang, X.-T.; Du, X.-Y.; Fang, Z.; Liu, Z.; Xu, Y.; Zheng, P.; Xu, X.-J.; Cheng, P.-F.; Huang, T.; et al. Severe disturbance of glucose metabolism in peripheral blood mononuclear cells of schizophrenia patients: A targeted metabolomic study. *J. Transl. Med.* **2015**, *13*, 226. [CrossRef]

60. Yang, X.; Sun, L.; Zhao, A.; Hu, X.; Qing, Y.; Jiang, J.; Yang, C.; Xu, T.; Wang, P.; Liu, J.; et al. Serum fatty acid patterns in patients with schizophrenia: A targeted metabonomics study. *Transl. Psychiatry* **2017**, *7*, e1176. [CrossRef]

61. Peters, B.D.; Machielsen, M.W.J.; Hoen, W.P.; Caan, M.W.A.; Malhotra, A.K.; Szeszko, P.R.; Duran, M.; Olabarriaga, S.D.; de Haan, L. Polyunsaturated fatty acid concentration predicts myelin integrity in early-phase psychosis. *Schizophr. Bull.* **2013**, *39*, 830–838. [CrossRef]

62. Fernstrom, J.D. Branched-chain amino acids and brain function. *J. Nutr.* **2005**, *135*, 1539S–1546S. [CrossRef]

63. Bevans, M.L.; Read, L.L.; Chao, H.M.; Clelland, J.D.; Suckow, R.F.; Maher, T.J.; Citrome, L.; Richardson, M.A. Efficacy of the Branched-Chain Amino Acids in the Treatment of Tardive Dyskinesia in Men. *Am. J. Psychiatry* **2003**, *160*, 1117–1124.

64. Włodarczyk, A.; Wiglusz, M.S.; Cubała, W.J. Ketogenic diet for schizophrenia: Nutritional approach to antipsychotic treatment. *Med. Hypotheses* **2018**, *118*, 74–77. [CrossRef] [PubMed]

65. Penttilä, M.; Jaaskelainen, E.; Hirvonen, N.; Isohanni, M.; Miettunen, J. Duration of untreated psychosis as predictor of long-term outcome in schizophrenia: Systematic review and meta-analysis. *Br. J. Psychiatry* **2014**, *205*, 88–94. [CrossRef] [PubMed]

66. Lin, C.-H.; Lin, S.-C.; Huang, Y.-H.; Wang, F.-C.; Huang, C.-J. Early prediction of olanzapine-induced weight gain for schizophrenia patients. *Psychiatry Res.* **2018**, *263*, 207–211. [CrossRef] [PubMed]

67. Castro-Nallar, E.; Bendall, M.L.; Pérez-Losada, M.; Sabuncyan, S.; Severance, E.G.; Dickerson, F.B.; Schroeder, J.R.; Yolken, R.H.; Crandall, K.A. Composition, taxonomy and functional diversity of the oropharynx microbiome in individuals with schizophrenia and controls. *PeerJ* **2015**, *3*, e1140. [CrossRef]

68. Palleja, A.; Mikkelsen, K.H.; Forslund, S.K.; Kashani, A.; Allin, K.H.; Nielsen, T.; Hansen, T.H.; Liang, S.; Feng, Q.; Zhang, C.; et al. Recovery of gut microbiota of healthy adults following antibiotic exposure. *Nat. Microbiol.* **2018**, *3*, 1255–1265. [CrossRef]

69. Bahr, S.M.; Tyler, B.C.; Wooldridge, N.; Butcher, B.D.; Burns, T.L.; Teesch, L.M.; Oltman, C.L.; Azcarate-Peril, M.A.; Kirby, J.R.; Calarge, C.A. Use of the second-generation antipsychotic, risperidone, and secondary weight gain are associated with an altered gut microbiota in children. *Transl. Psychiatry* **2015**, *5*, e652. [CrossRef]

70. Marlicz, W.; Yung, D.E.; Skonieczna-Żydecka, K.; Loniewski, I.; Van Hemert, S.; Loniewska, B.; Koulaouzidis, A. From clinical uncertainties to precision medicine: The emerging role of the gut barrier and microbiome in small bowel functional diseases. *Expert Rev. Gastroenterol. Hepatol.* **2017**, *11*, 961–978. [CrossRef]

71. Skonieczna-Żydecka, K.; Łoniewski, I.; Misera, A.; Stachowska, E.; Maciejewska, D.; Marlicz, W.; Galling, B. Second-generation antipsychotics and metabolism alterations: A systematic review of the role of the gut microbiome. *Psychopharmacology* **2018**, *236*, 1491–1512. [CrossRef]

72. Rao, T.S.S.; Asha, M.R.; Ramesh, B.N.; Rao, K.S.J. Understanding nutrition, depression and mental illnesses. *Indian J. Psychiatry* **2008**, *50*, 77–82.

73. Wu, G.D.; Chen, J.; Hoffmann, C.; Bittinger, K.; Chen, Y.-Y.; Keilbaugh, S.A.; Bewtra, M.; Knights, D.; Walters, W.A.; Knight, R.; et al. Linking long-term dietary patterns with gut microbial enterotypes. *Science* **2011**, *334*, 105–108. [CrossRef] [PubMed]

Journal of
Clinical Medicine

Article

Hydrogen Sulfide as a Toxic Product in the Small–Large Intestine Axis and its Role in IBD Development

Ivan Kushkevych [1],*, Dani Dordević [2], Peter Kollar [3], Monika Vítězová [1] and Lorenzo Drago [4]

1. Department of Experimental Biology, Faculty of Science, Masaryk University, Kamenice 753/5, 62500 Brno, Czech Republic
2. Department of Plant Origin Foodstuffs Hygiene and Technology, Faculty of Veterinary Hygiene and Ecology, University of Veterinary and Pharmaceutical Sciences, 61242 Brno, Czech Republic
3. Department of Human Pharmacology and Toxicology, Faculty of Pharmacy, University of Veterinary and Pharmaceutical Sciences, 61242 Brno, Czech Republic
4. Department of Biomedical Sciences for Health, University of Milan, 20122 Milan, Italy
* Correspondence: kushkevych@mail.muni.cz; Tel.: +420-549-495-315

Received: 2 July 2019; Accepted: 17 July 2019; Published: 19 July 2019

Abstract: The small–large intestine axis in hydrogen sulfide accumulation and testing of sulfate and lactate in the gut–gut axis of the intestinal environment has not been well described. Sulfate reducing bacteria (SRB) of the *Desulfovibrio* genus reduce sulfate to hydrogen sulfide and can be involved in ulcerative colitis development. The background of the research was to find correlations between hydrogen sulfide production under the effect of an electron acceptor (sulfate) and donor (lactate) at different concentrations and *Desulfovibrio piger* Vib-7 growth, as well as their dissimilatory sulfate reduction in the intestinal small–large intestinal environment. Methods: Microbiological, biochemical, and biophysical methods, and statistical processing of the results (principal component and cross-correlation analyses) were used. Results: *D. piger* Vib-7 showed increased intensity of bacterial growth and hydrogen sulfide production under the following concentrations of sulfate and lactate: 17.4 mM and 35.6 mM, respectively. The study showed in what kind of intestinal environment *D. piger* Vib-7 grows at the highest level and produces the highest amount of hydrogen sulfide. Conclusions: The optimum intestinal environment of *D. piger* Vib-7 can serve as a good indicator of the occurrence of inflammatory bowel diseases; meaning that these findings can be broadly used in medicine practice dealing with the monitoring and diagnosis of intestinal ailments.

Keywords: small–large intestine axis; hydrogen sulfide; *Desulfovibrio*; bowel disease; colitis

---

1. Introduction

The destination of food remains from the small intestine, together with microbial biomass, is the large intestine, which represents an open system of the small–large intestine axis [1]. This means that the large intestine is a reactor for constant microorganism cultivation [2]. This fact is supported by the calculation that 200 g of digestive material is present in the large intestine of an adult human [2,3]. The intestinal lumen biomass includes almost 55% microorganisms, which are present in the total fecal content [1,4,5]. The microbial mass in the large intestine is 10^{11}–10^{12} cells/g feces of the following dominant genera: *Bifidobacterium*, *Bacteroides*, *Lactobacillus*, *Escherichia*, *Enterococcus*, *Atopobium*, *Faecalibacterium*, *Clostridium*, and 40 other bacterial species that represent 99% of the colon microbiota [1,4,6,7].

The majority of these bacteria are able to cleave complex organic compounds in the fermentation process and they produce molecular hydrogen, different acids including acetate and lactate, and other compounds. The production of lactate depends on the fermentative properties of lactic acid bacteria (e.g., *Bifidobacterium, Lactobacillus,* and *Streptococcus*) [4]. This means that lactate and acetate can be also used by other groups of microorganisms. These compounds can be used as electron donors and carbon sources in the metabolic processes of microorganisms [7–10]. Intestinal microbiota is especially involved in the digestion processes of short-chain fatty acids [4]. The physiology and metabolism of humans is highly dependent on intestinal microorganisms and consequently affects human physiological functions and health [1–3,11,12]. On the other hand, another important component of human physiological status is the capability of the intestinal system to absorb sulfate for amino acid development, such as cysteine and methionine, and its regular involvement in assimilation processes. Concentrations of sulfate in the intestine are dependent on human diet since sulfate is present differently in different food commodities [13–16]; another factor is that sulfate absorption is done individually in each human, meaning that the total sulfate content in the intestine is highly influenced by eating habits. The importance of daily sulfate intake can be emphasized by the fact that staple food commodities (such as some breads) represent high sulfate sources (>10 μmol/g) as do popular beverages such as beers and wines (>2.5 μmol/g) [13].

The remnants of sulfate that are not absorbed by the intestines and the presence of lactate make a good environment for sulfate-reducing bacteria (SRB) that are regularly found in human and animal intestines [1,4,17–21]. SRB use sulfate as the final electron acceptor in the process of dissimilatory sulfate reduction and form the end product of hydrogen sulfide [22–27]. Different organic compounds, including lactate, can be exogenic electron donors for this process and can be oxidized to acetate [18,28]. *Desulfovibrio* genus is the dominant SRB in the human intestine [5,22]. Previous studies indicated a correlation between the SRB intestinal presence and ailments, such as cholecystitis, brain abscesses, and abdominal cavity ulcerative enterocolitis, making *Desulfovibrio* species an important factor during both mono- and poly-microbial infections of the gastrointestinal tract [2–4,12]. Consequently, the prevalence of SRB in the intestines is dependent on the occurrence of sulfate and lactate presence in the gut. It is also important to stress that the intestinal environment should be monitored due to its influence on SRB since a connection with these bacterial strains and inflammatory bowel diseases (IBD) has been found [1–3,11]. The effects of sulfate and lactate at different concentrations on intestinal *Desulfovibrio* species growth and their sulfate reduction parameters have not been well studied.

The aim of this research was to find correlations between different sulfate and lactate concentrations and *Desulfovibrio piger* Vib-7 growth parameters and their dissimilatory sulfate reduction in the small–large intestinal environment.

2. Experimental Section

2.1. Bacterial Culture and Cultivation

The sulfate-reducing bacteria of the *Desulfovibrio piger* strain Vib-7 was used as the object of the study. This strain was isolated from the human large intestine and identified based on physiological and biochemical properties and sequence analysis of the 16S rRNA gene. The accession number in GenBank is KT881309.1. The strain of SRB was kept in the collection of microorganisms at the Laboratory of Anaerobic Microorganisms of the Department of Experimental Biology at Masaryk University (Brno, Czech Republic).

The bacterial culture was grown in modified liquid Postgate C medium [23] for 72 h at 37 °C under anaerobic conditions [29]. The following sodium sulfate concentrations were prepared in medium: 0.87 mM, 1.75 mM, 3.5 mM, 7 mM, 10.5 mM, and 17.5 mM. Different concentrations of electron donors and their effect in the medium were determined in the presence of sodium lactate (4.45 mM, 8.9 mM, 17.8 mM, 35.6 mM, 53.4 mM, or 89 mM). The control medium consisted of 3.5 mM sulfate and 17.8 mM lactate. The determination of biomass, sulfate, hydrogen sulfide, lactate, and acetate concentrations were determined after 12, 24, 36, 48, 60, and 72 h.

2.2. Bacterial Biomass Determination

In total, 1 mL of liquid medium without Mohr's salt in a plastic cuvette was measured in a biophotometer (Eppendorf BioPhotometer®D30, Hamburg, Germany) as a blank. The same procedure with the bacterial suspension was performed. The optical density (OD_{340}) was always measured before the experiments to provide approximately the same amount of SRB in each experiment [7].

2.3. Sulfate Determination

The sulfate concentration in the liquid medium was measured by turbidimetric method after 12 h intervals of cultivation. In total, 40 mg/L $BaCl_2$ solution was prepared in 0.12 M HCl and mixed with glycerol in a 1:1 ratio. The supernatant of the sample was obtained by centrifugation at 5000× g at 23 °C and 1 mL was added to 10 mL of $BaCl_2$:glycerol solution and carefully mixed. The absorbance of the mixed solution was measured after 10 min at 520 nm (Spectrosonic Genesis 5, Ecublens, Switzerland). A cultivation medium without bacteria growth was used as a control [30].

2.4. Hydrogen Sulfide Determination

The concentration of hydrogen sulfide was determined in cultivation medium after different time intervals. In total, 1 mL of the sample was added to 10 mL of a 5 g/L solution of zinc acetate and 2 mL of 0.75 g/mL p-aminodimethylaniline in a solution of sulfuric acid (2 M). The mixture stood for 5 min at room temperature. After that, 0.5 mL of 12 g/L solution of ferric chloride dissolved in 15 mM sulfuric acid was added. After standing another 5 min at room temperature, the mixture was centrifuged 5000× g at 23 °C. The absorbance of the mixture was determined at a wavelength of 665 nm by a spectrophotometer (Cecil Aquarius CE 7200 Double Beam Spectrophotometer, London, UK) [31,32].

2.5. Lactate and Acetate Determination

The measurement was repeated in the same manner using a cultivation medium and it served as the control sample. Measurements of lactate concentration using a lactate assay kit (Sigma-Aldrich, Catalog Number MAK064, Prague, Czech Republic) were carried out. Accumulation of acetate ions in the process of bacterial growth in the medium was determined using the acetate assay kit (Abnova, Colorimetric, Catalog Number KA3764, Prague, Czech Republic).

2.6. Statistical Analysis

Using the experimental data, the basic statistical parameters (M—mean, m—standard error, M ± m) were calculated. The accurate approximation was when $p \leq 0.0533$ [33]. Statistical significance was measured with the use of principal component analysis (PCA) that gave overall differences among compared groups. Statistical analysis was done by SPSS 20 statistical software (IBM Corporation, Armonk, NY, USA). Plots were built by software package Origin7.0 (Northampton, UK).

3. Results

Intestinal sulfate-reducing bacteria, *D. piger* Vib-7, showed the highest rate (biomass accumulation, sulfate and lactate consumption, and sulfide and acetate production), both increasing and decreasing trends, until the 60th h of cultivation in the control (3.5 mM of sulfate and 17.3 mM of lactate) medium (Figure 1). The stationary growth phase was achieved after 60 h of cultivation and the following percentage decreases and increases in contents were measured: biomass (increased by 87%), sulfate (decreased by 95%), sulfide (increased by 83%), lactate (decreased by 88%) and acetate (increased by 91%). Relative growth and survival of *D. piger* Vib-7 achieved the highest percentages at 7 mM of sulfate and 35.6 mM of lactate. Higher concentrations than these resulted in the stability of relative growth and it stayed at the same level during 12 to 48 h. Lower concentrations of sulfate (<3.5 mM) and lactate (<17.8 mM) were not enough for the achievement of maximum growth parameters.

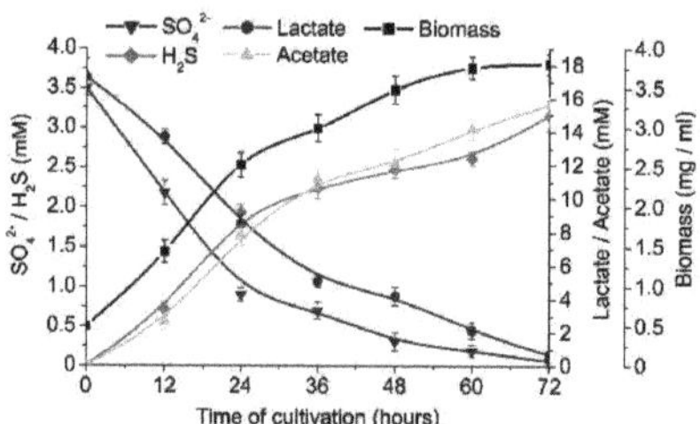

Figure 1. The growth of *D. piger* Vib-7 and their sulfate reduction.

As can be seen in Figure 2, the consumption of sulfate was highly dependent on its different concentrations in cultivation medium, time of cultivation, and the presence of lactate donors (it was constant at 17.8 mM of lactate).

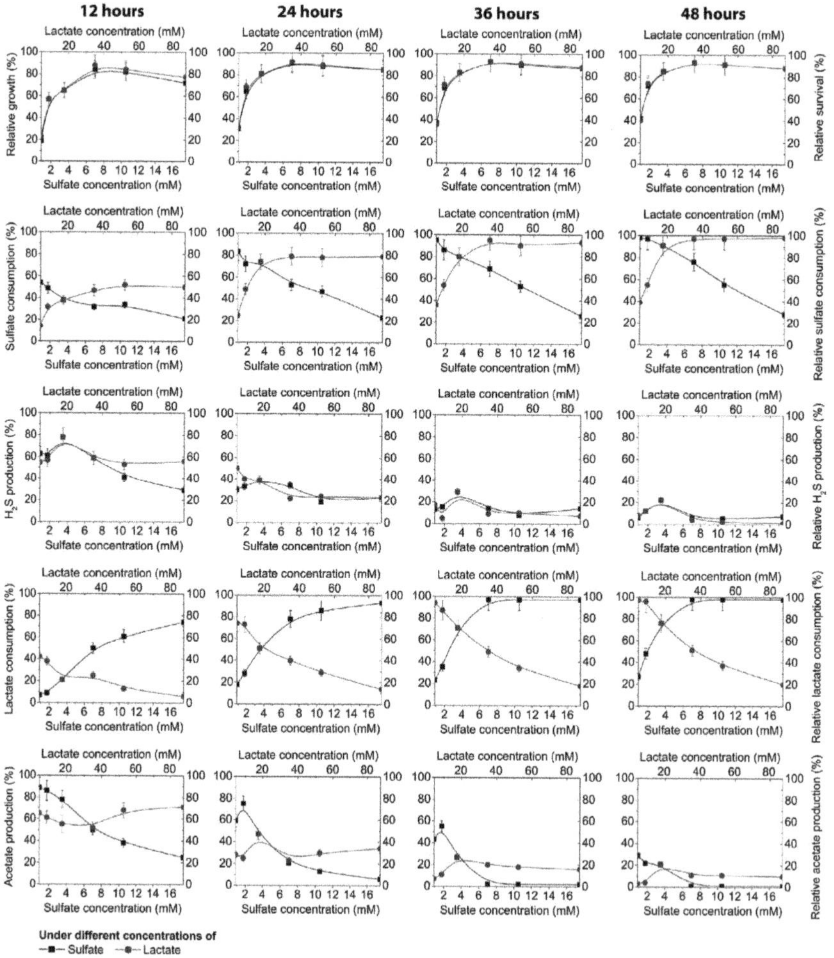

Figure 2. Growth of *D. piger* Vib-7, and their survival and sulfate reduction parameters during 12, 24, 36, and 48 h of cultivation: the effect of electron acceptor (sulfate)/donor (lactate) at different concentrations (columns: first = 12 h, second = 24 h, third = 36 h, fourth = 48 h).

After 12 h, 54% of the sulfate was consumed in medium with lower sulfate concentrations (0.87 mM), although after 48 h, sulfate was almost consumed (98%) at the lowest concentration (0.87 mM) and only 28% at the highest sulfate concentration, where 72% was not used during this time period. Under other conditions, the following changes occurred: different lactate concentrations (4.45 mM, 8.9 mM, 17.8 mM, 35.6 mM, 53.4 mM, or 89 mM) were added in the cultivation medium and the consumption of the sulfate was measured.

As can be seen in Figure 2, sulfate consumption depended not only on its concentration, but was also strongly correlated with the concentration of an electron donor (lactate). Within this environment 14% of the sulfate was used at the lowest lactate concentration (4.45 mM) and 50% at 89 mM of lactate in the medium after 12 h of cultivation. The time of cultivation and lactate concentration increased the sulfate reduction in the medium. After 48 h, sulfate was used only 39–55% at the lowest concentrations of lactate (4.45–8.9 mM) because not enough electron donor was present. However, increasing the lactate concentrations from 35.6 to 89 mM induced 91–98% consumption of sulfate. The same trend was noticed with the lactate consumption. It could be seen that the production of sulfide was not very much influenced by the concentration of electron acceptor (0.87 mM to 17.5 mM), or the electron donor (4.45 mM to 89 mM), in the time interval from 24 to 48 h. The hydrogen sulfide production during this time period was stable. The highest production (78%) of sulfide was accumulated during the first 12 h and gradually decreased to 39%, 29%, and 22%, after 24, 36, and 48 h, respectively, under the conditions of 3.5 mM sulfate and 17.8 mM lactate. A similar trend was noticed in acetate production, although acetate production was more influenced by the sulfate and lactate concentration in the medium, as well as by the cultivation time. The highest production of acetate was until the 36th h of cultivation and after this period it decreased (Figure 2).

Based on different concentrations of electron acceptor and donor, PCA was carried out (Figure 3) that included the separate parameters of biomass, sulfate and lactate consumption, and H_2S and acetate production, as well as PCA that included all mentioned parameters.

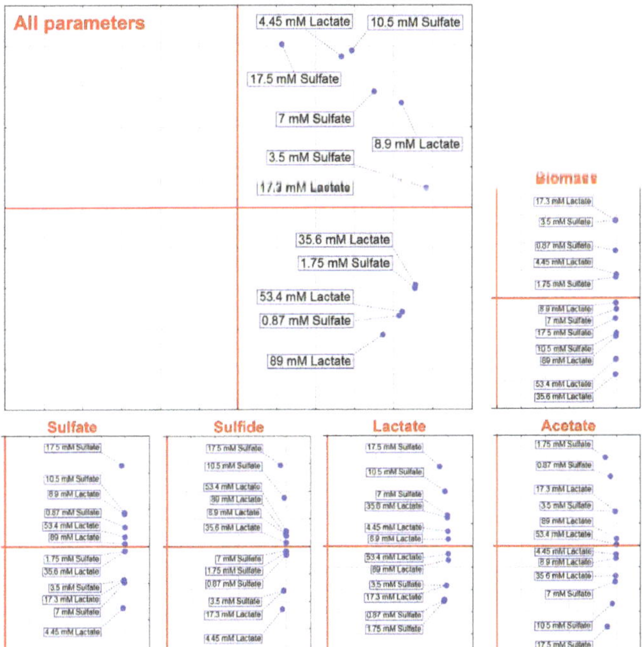

Figure 3. Principal component analysis of the *D. piger* Vib-7 growth and the parameters of sulfate reduction under the effect of electron acceptor (sulfate)/donor (lactate) at different concentrations.

PCA that included separate parameters did not show clusters that would indicate a trend observed in Figure 2, but PCA that included all parameters showed that concentrations of 53 mM lactate and 0.87 mM sulfate, 1.75 mM sulfate and 35.6 mM lactate, and 3.5 mM sulfate and 17.3 mM lactate formed separated clusters. These findings indicated that lower concentrations of sulfate were prevailing in an environment with higher concentrations of lactate.

To observe side shifts in the process of sulfate reduction in the intestinal environment, including different concentrations of sulfate and lactate, cross correlation analysis was carried out between the following parameters: biomass and sulfate, biomass and sulfide, biomass and lactate, biomass and acetate, sulfate and sulfide, sulfate and lactate, sulfate and acetate, sulfide and lactate, sulfide and acetate, and lactate and acetate (Figure 4).

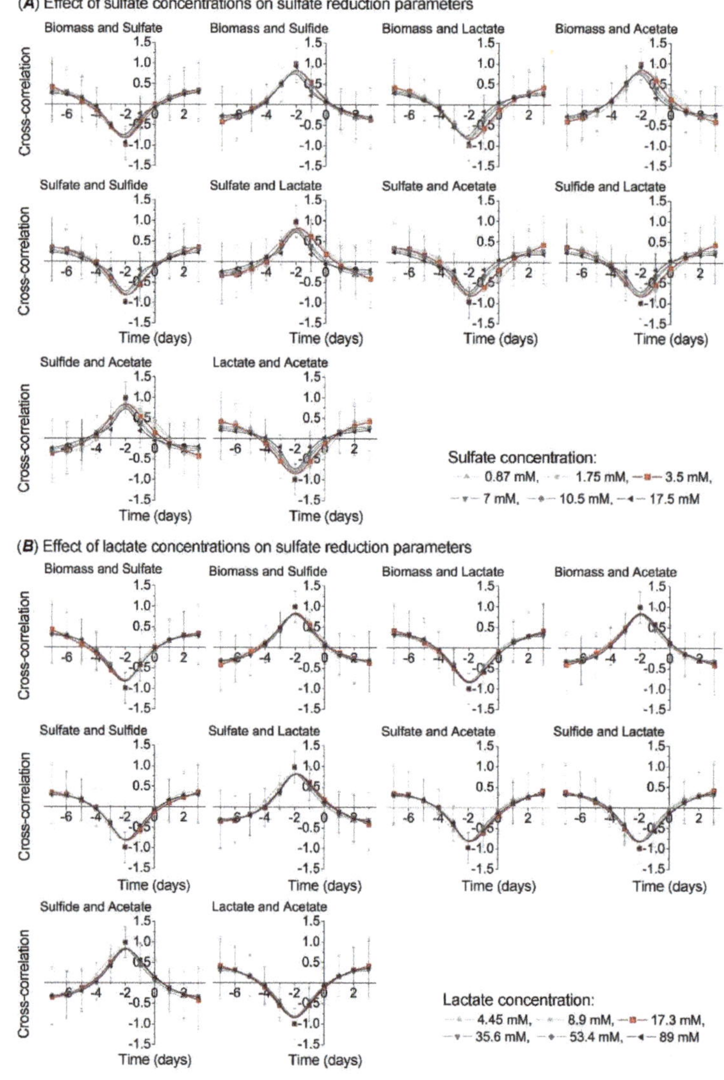

Figure 4. Cross-correlation analysis between growth (biomass) and sulfate reduction parameters under the effect of electron acceptor (sulfate)/donor (lactate) at different concentrations.

The higher sulfate concentrations resulted in a shift to the left or right side on the Y axis, in comparison to the control sample (red line in Figure 4A), by all parameters, although more significantly by the following parameters: biomass and lactate, biomass and acetate, sulfate and lactate, and sulfate and acetate. Oppositely, lactate concentration effect did not cause similar shifting on the Y axis (Figure 4B).

PCA of the *D. piger* Vib-7 growth and the parameters of sulfate reduction based on cross-correlation analysis clearly showed an isolated cluster of the highest sulfate consumption (17.5 mM) in comparison with other concentrations. This means that bacteria were not able to fully consume these high sulfate concentrations during 48 h of cultivation (Figure 5).

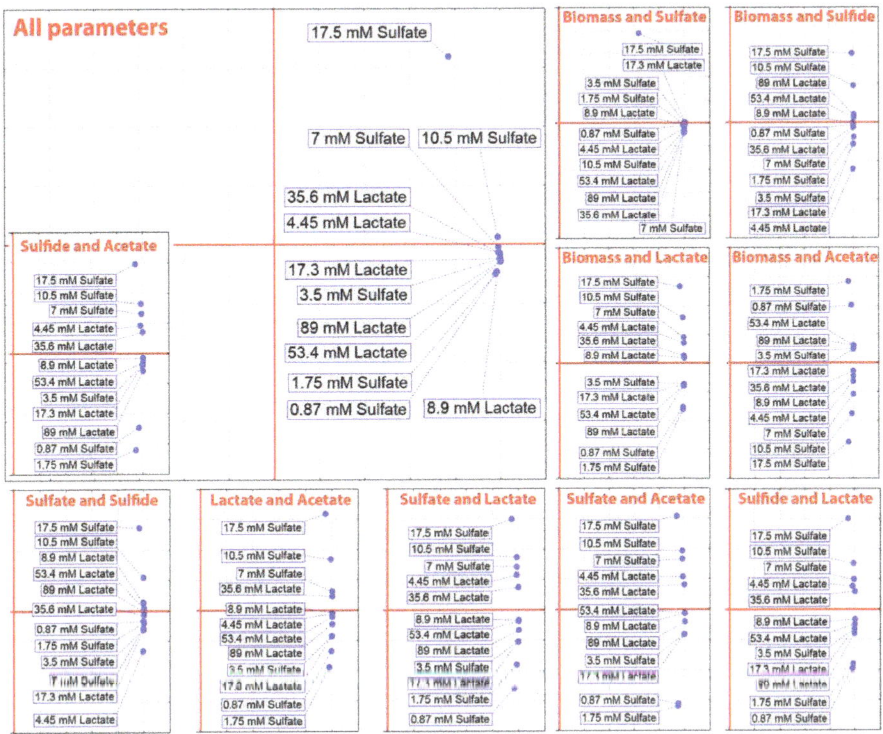

Figure 5. Principal component analysis of the *D. piger* Vib-7 growth and the parameters of sulfate reduction based on cross-correlation analysis.

The kinetic parameters of *D. piger* Vib-7 growth under the effect of electron acceptor (sulfate)/donor (lactate) at different concentrations are shown in Table 1.

Table 1. Kinetics of *D. piger* Vib-7 growth under the effect of electron acceptor/donor.

Electron Acceptor (Sulfate)				Electron Donor (Lactate)			
Sulfate (mM)	Lag-Phase (h)	Generation Time T_d (h)	μ_{max} (h^{-1})	Lactate (mM)	Lag-Phase (h)	Generation Time T_d (h)	μ_{max} (h^{-1})
0.87	38.2 ± 3.5	16.5 ± 1.5	0.009 ± 0.0001	4.45	36.6 ± 3.7	14.5 ± 1.35	0.009 ± 0.008
1.75	5.9 ± 0.46	4.3 ± 0.44	0.02 ± 0.001	8.9	7.1 ± 0.66	3.6 ± 0.33	0.03 ± 0.001
3.5	6.4 ± 0.62	1.8 ± 0.15	0.05 ± 0.004	17.3	6.4 ± 0.60	1.8 ± 0.12	0.05 ± 0.004
7.0	7.4 ± 0.73	1.1 ± 0.10	0.08 ± 0.007	35.6	4.9 ± 0.43	1.1 ± 0.10	0.08 ± 0.007
10.5	3.3 ± 0.31	1.3 ± 0.12	0.06 ± 0.005	53.4	3.1 ± 0.29	1.3 ± 0.11	0.07 ± 0.005
17.5	5.5 ± 0.59	1.6 ± 0.14	0.05 ± 0.005	89.0	5.4 ± 0.51	1.5 ± 0.13	0.06 ± 0.004

Under sulfate concentrations of 10.5 mM the shortest lag phase was measured and specific maximum rate of growth (μmax) was the fastest at 7.0 mM of sulfate. Under electron donor (lactate) concentrations the shortest lag phase and the fastest specific maximum rate of growth were detected at 53.4 mM and 35.6 mM of lactate, respectively.

4. Discussion

The sulfate consumption and sulfide production, and the lactate consumption and acetate accumulation are important factors influencing the intestinal environment [7–10]. Intestinal sulfate-reducing bacteria, especially *Desulfovibrio* genus, are often found in the intestines and feces of people and animals with IBD. One of the main roles in the development of colitis, among other factors, can also be the species of this genus. These bacteria use sulfate as a terminal electron acceptor and organic compounds as electron donors in their metabolism [6,7]. This fact leads us to the conclusion that sulfate present in the daily diet plays an important role in the development of bowel disease. Sulfate is present mainly in the following food commodities: some breads, soya flour, dried fruits, brassicas, and sausages, as well as some beers, ciders, and wines. These data indicate that sulfate intake is highly dependent on diet and the small–large intestine axis [13].

In our previous research, principal component analysis indicated that the *Desulfovibrio* strains from individuals with colitis were grouped in one cluster by biomass accumulation and sulfide production, and the strains from healthy individuals formed another cluster by the same parameters. Sulfate and lactate consumption measured over time showed a negative correlation (Pearson correlations, $p < 0.01$). The linear regression (R^2) was lower in biomass accumulation and hydrogen sulfide production. Thus, biomass accumulation and sulfide production, together with measured kinetic parameters, play an important factor in bowel inflammation, including ulcerative colitis. Additionally, acetate produced by SRB can also be in synergic interaction with H_2S, while sulfate consumption and lactate oxidation likely represent minor factors in bowel disease [16].

Our results provide an opportunity to find the optimum growing point of the bacteria. The study confirmed an intense growth of *D. piger* Vib-7 in the presence of higher concentrations of electron acceptor and donor, though the consequence is an intensive accumulation of sulfide and acetate. Data from the literature indicate that these conditions can be the cause of ulcerative colitis that can lead to cancer of the bowel. This statement is supported by the fact that hydrogen sulfide negatively affects intestinal mucosa and epithelial cells, inhibits the growth of colonocytes [4,14–18,34–37], causes phagocytosis, causes the death of intestinal bacteria [4,12,24], and induces hyperproliferation and metabolic abnormalities of epithelial cells [12]. The high level of metabolites and the presence of SRB are connected with the inflammation of the colon [4,6,36]. Therefore, the integrity of colonocytes is maintained by hydrogen sulfide concentration [35–37]. Sulfide production is higher among SRB isolated from individuals with ulcerative colitis [5,6].

Other research describing cross-correlation parameters of the SRB metabolic process indicated that the strains isolated from people with colitis shifted to the right side of the Y axis by biomass accumulation, sulfate consumption, lactate oxidation, as well as hydrogen sulfide and acetate production, compared with the strains isolated from healthy individuals. Different percentages were observed in shifting to the right side of the Y axis: biomass accumulation 26%, sulfate consumption 1.5%, and sulfide production 5% [14]. It should be noted that the intestinal microbiota is a very complex system that may limit this study. There are a lot of interactions with clostridia, methanogens, lactic acid bacteria, etc. However, a central role in the development of IBD, especially ulcerative colitis, is SRB [1–3,11]. This bacterial group, producing hydrogen sulfide, can inhibit other microbiota, including lactic acid bacteria, methanogens, and many other intestinal microorganisms [2].

A diet high in sulfate ions (preservatives added to food often contain sulfur oxides) causes an increase in hydrogen sulfide concentration by SRB in rumens. The studies have revealed that the western diet contains over 16.6 mmol sulfate/day [13] and the feces of approximately 50% of healthy individuals contain SRB (up to 92% belong to the genus *Desulfovibrio*) [1,5]. Sulfate polysaccharides such as mucin, chondroitin sulfate, and carrageenan are broadly consumed, and they represent good sources of sulfate for SRB [24]. It should also be noted that hydrogen sulfide can be toxic not only for intestinal cells, but also for its producers. The highest toxicity of H_2S was measured in the presence of concentrations higher than 6 mM, where growth was stopped, though metabolic activities were not 100% inhibited. These findings are confirmed by cross correlation and principal component analysis that clearly support the above mentioned results. The presence of 5 mM H_2S resulted in a two times longer lag phase and generation time was eight times longer. The results confirmed toxicity of H_2S toward *Desulfovibrio* [18]. Beside sulfate and lactate, terminal oxidative processes in the human large intestine could be involved in the activities of SRB, and consequently the production of hydrogen sulfide in high concentrations that can cause inflammatory bowel disease development.

5. Conclusions

The study gave more information about the intestinal environment in vitro concerning sulfate and lactate concentrations and their effects on the growth parameters of *Desulfovibrio piger* Vib-7. Almost total consumption of sulfate and lactate was achieved after 60 h of cultivation, though the best relative growth and stability was measured at 7 mM and 35.6 mM of sulfate and lactate, respectively. PCA including separated parameters did not show combined clusters, but PCA based on all parameters showed that different concentrations of sulfate and lactate formed separated clusters. These obtained results represent the main findings of the research, indicating that SRB would grow at the highest level under these experimentally simulated conditions. These conditions are an indicator of higher SRB activity that can lead to the development of IBD, and further studies will certainly focus more on the intestinal environment concerning SRB not only in vitro, but also in vivo.

Author Contributions: Conceptualization, I.K. and D.D.; methodology, I.K. and M.V.; validation, P.K., M.V., and D.D.; formal analysis, M.V. and L.D.; investigation, I.K.; resources, I.K.; data curation, D.D.; writing—original draft preparation, I.K., D.D., and M.V.; writing—review and editing, I.K. and L.D.; visualization, I.K.; supervision, M.V.; project administration, I.K.; funding acquisition, I.K., D.D., and P.K.

Funding: This research was supported by Grant Agency of the Masaryk University (MUNI/A/0902/2018).

References

1. Gibson, G.R.; Cummings, J.H.; Macfarlane, G.T. Growth and activities of sulphate-reducing bacteria in gut contents of health subjects and patients with ulcerative colitis. *FEMS Microbiol. Ecol.* **1991**, *86*, 103–112. [CrossRef]

2. Gibson, G.R.; Macfarlane, S.; Macfarlane, G.T. Metabolic interactions involving sulphate-reducing and methanogenic bacteria in the human large intestine. *FEMS Microbiol. Ecol.* **1993**, *12*, 117–125. [CrossRef]

3. Cummings, J.H.; Macfarlane, G.T.; Macfarlane, S. Intestinal Bacteria and Ulcerative Colitis. *Curr. Issues Intest. Microbiol.* **2003**, *4*, 9–20. [PubMed]

4. Barton, L.L.; Hamilton, W.A. *Sulphate-Reducing Bacteria Environmental and Engineered Systems*; Cambridge University Press: Cambridge, UK, 2017.

5. Loubinoux, J.; Bronowicji, J.P.; Pereira, I.A. Sulphate-reducing bacteria in human feces and their association with inflammatory diseases. *FEMS Microbiol. Ecol.* **2002**, *40*, 107–112. [CrossRef] [PubMed]

6. Ková č, J.; Vítězová, M.; Kushkevych, I. Metabolic activity of sulfate-reducing bacteria from rodents with colitis. *Open Med.* **2018**, *13*, 344–349.

7. Kushkevych, I.; Vítězová, M.; Fedrová, P.; Vochyanová, Z.; Paráková, L.; Hošek, J. Kinetic properties of growth of intestinal sulphate-reducing bacteria isolated from healthy mice and mice with ulcerative colitis. *Acta Vet. Brno* **2017**, *86*, 405–411. [CrossRef]

8. Kushkevych, I.; Fafula, R.; Parak, T.; Bartoš, M. Activity of Na^+/K^+-activated Mg^{2+}-dependent ATP hydrolase in the cell-free extracts of the sulfate-reducing bacteria Desulfovibrio piger Vib-7 and Desulfomicrobium sp. Rod-9. *Acta Vet. Brno* **2015**, *84*, 3–12. [CrossRef]

9. Kushkevych, I.V. Activity and kinetic properties of phosphotransacetylase from intestinal sulfate-reducing bacteria. *Acta Biochem. Pol.* **2015**, *62*, 1037–1108. [CrossRef]

10. Kushkevych, I.V. Kinetic Properties of Pyruvate Ferredoxin Oxidoreductase of Intestinal Sulfate-Reducing Bacteria Desulfovibrio piger Vib-7 and Desulfomicrobium sp. Rod-9. *Pol. J. Microbiol.* **2015**, *64*, 107–114.

11. Loubinoux, J.; Mory, F.; Pereira, I.A.; Le Faou, A.E. Bacteremia caused by a strain of Desulfovibrio related to the provisionally named Desulfovibrio fairfieldensis. *J. Clin. Microbiol.* **2000**, *38*, 931–934.

12. Pitcher, M.C.; Cummings, J.H. Hydrogen sulphide: A bacterial toxin in ulcerative colitis? *Gut* **1996**, *39*, 1–4. [CrossRef] [PubMed]

13. Florin, T.H.; Neale, G.; Goretski, S. Sulfate in food and beverages. *J. Food Compos. Anal.* **1993**, *6*, 140–151. [CrossRef]

14. Kushkevych, I.; Dordević, D.; Vítězová, M.; Kollár, P. Cross-correlation analysis of the Desulfovibrio growth parameters of intestinal species isolated from people with colitis. *Biologia* **2018**, *73*, 1137–1143. [CrossRef]

15. Kushkevych, I.; Dordević, D.; Vítězová, M. Analysis of pH dose-dependent growth of sulfate-reducing bacteria. *Open Med.* **2019**, *14*, 66–74. [CrossRef] [PubMed]

16. Kushkevych, I.; Dordević, D.; Kollar, P. Analysis of physiological parameters of Desulfovibrio strains from individuals with colitis. *Open Life Sci.* **2018**, *13*, 481–488. [CrossRef]

17. Kushkevych, I.; Vítězová, M.; Kos, J.; Kollár, P.; Jampilek, J. Effect of selected 8-hydroxyquinoline-2-carboxanilides on viability and sulfate metabolism of Desulfovibrio piger. *J. Appl. Biomed.* **2018**, *16*, 241–246. [CrossRef]

18. Kushkevych, I.; Dordević, D.; Vítězová, M. Toxicity of hydrogen sulfide toward sulfate-reducing bacteria Desulfovibrio piger Vib-7. *Arch. Microbiol.* **2019**, *201*, 1–9. [CrossRef] [PubMed]

19. Kushkevych, I.; Kollar, P.; Suchy, P.; Parak, T.; Pauk, K.; Imramovsky, A. Activity of selected salicylamides against intestinal sulfate-reducing bacteria. *Neuroendocrinol. Lett.* **2015**, *36*, 106–113. [PubMed]

20. Kushkevych, I.; Kollar, P.; Ferreira, A.L.; Palma, D.; Duarte, A.; Lopes, M.M.; Bartos, M.; Pauk, K.; Imramovsky, A.; Jampilek, J. Antimicrobial effect of salicylamide derivatives against intestinal sulfate-reducing bacteria. *J. Appl. Biomed.* **2016**, *14*, 125–130. [CrossRef]

21. Kushkevych, I.; Kos, J.; Kollar, P.; Kralova, K.; Jampilek, J. Activity of ring-substituted 8-hydroxyquinoline-2-carboxanilides against intestinal sulfate-reducing bacteria Desulfovibrio piger. *Med. Chem. Res.* **2018**, *27*, 278–284. [CrossRef]

22. Loubinoux, J.; Valente, F.M.A.; Pereira, I.A.C. Reclassification of the only species of the genus Desulfomonas, Desulfomonas pigra, as Desulfovibrio piger comb. nov. *Int. J. Syst. Evol. Microbiol.* **2002**, *52*, 1305–1308. [PubMed]

23. Postgate, J.R. *The Sulfate Reducing Bacteria*; Cambridge University Press: Cambridge, UK, 1984.

24. Rowan, F.E.; Docherty, N.G.; Coffey, J.C.; O'Connell, P.R. Sulphate-reducing bacteria and hydrogen sulphide in the aetiology of ulcerative colitis. *Br. J. Surg.* **2009**, *96*, 151–158. [CrossRef] [PubMed]

25. Kushkevych, I.; Vítězová, M.; Vítěz, T.; Bartoš, M. Production of biogas: Relationship between methanogenic and sulfate-reducing microorganisms. *Open Life Sci.* **2017**, *12*, 82–91. [CrossRef]

26. Kushkevych, I.; Vítězová, M.; Vítěz, T.; Kovac, J.; Kaucká, P.; Jesionek, W.; Bartoš, M.; Barton, L. A new combination of substrates: Biogas production and diversity of the methanogenic microorganisms. *Open Life Sci.* **2018**, *13*, 119–128. [CrossRef]

27. Kushkevych, I.; Kováč, J.; Vítězová, M.; Vítěz, T.; Bartoš, M. The diversity of sulfate-reducing bacteria in the seven bioreactors. *Arch. Microbiol.* **2018**, *200*, 945–950. [CrossRef] [PubMed]

28. Černý, M.; Vítězová, M.; Vítěz, T.; Bartoš, M.; Kushkevych, I. Variation in the Distribution of Hydrogen Producers from the Clostridiales Order in Biogas Reactors Depending on Different Input Substrates. *Energies* **2018**, *11*, 3270.

29. Kováč, J.; Kushkevych, I. New modification of cultivation medium for isolation and growth of intestinal sulfate-reducing bacteria. In Proceedings of the International PhD Students Conference MendelNet, Brno, Czechia, 6–7 November 2019; pp. 702–707.

30. Kolmert, A.; Wikstrom, P.; Hallberg, K.B. A fast and simple turbidimetric method for the determination of sulfate in sulfate-reducing bacterial cultures. *J. Microbiol. Methods* **2000**, *41*, 179–184. [CrossRef]

31. Cline, J.D. Spectrophotometric determination of hydrogen sulfide in natural water. *Limnol. Oceanogr.* **1969**, *14*, 454–458. [CrossRef]

32. Bailey, T.S.; Pluth, M.D. Chemiluminescent detection of enzymatically produced hydrogen sulfide: Substrate hydrogen bonding influences selectivity for H$_2$S over biological thiols. *J. Am. Chem. Soc.* **2013**, *135*, 16697–16704. [CrossRef]

33. Bailey, N.T.J. *Statistical Methods in Biology*; Cambridge University Press: Cambridge, UK, 1995.

34. Attene-Ramos, M.S.; Wagner, E.D.; Plewa, M.J.; Gaskins, H.R. Evidence that hydrogen sulfide is a genotoxic agent. *Mol. Cancer Res.* **2006**, *4*, 9–14. [CrossRef]

35. Beauchamp, R.O.; Bus, J.S.; Popp, J.A.; Boreiko, C.J.; Andjelkovich, D.A.; Leber, P. A critical review of the literature on hydrogen sulfide toxicity. *CRC Crit. Rev. Toxicol.* **1984**, *13*, 25–97. [CrossRef] [PubMed]

36. Blachier, F.; Davila, A.M.; Mimoun, S. Luminal sulfide and large intestine mucosa: Friend or foe? *Amino Acids* **2010**, *39*, 335–347. [CrossRef] [PubMed]

37. Grieshaber, M.K.; Völkel, S. Animal adaptations for tolerance and exploitation of poisonous sulfide. *Annu. Rev. Physiol.* **1998**, *60*, 33–53. [CrossRef] [PubMed]

Journal of
Clinical Medicine

Review

Upper Respiratory Tract Microbiome and Otitis Media Intertalk: Lessons from the Literature

Francesco Folino [1,*], Luca Ruggiero [2], Pasquale Capaccio [3,4], Ilaria Coro [1,2], Stefano Aliberti [1,5], Lorenzo Drago [6], Paola Marchisio [1,2] and Sara Torretta [3,7]

1. Department of Pathophysiology and Transplantation, University of Milan, 20122 Milan, Italy; ilaria.coro@policlinico.mi.it (I.C.); stefano.aliberti@unimi.it (S.A.); paola.marchisio@unimi.it (P.M.)
2. Pediatric Highly Intensive Care Unit, Fondazione IRCCS Ca' Granda Ospedale Maggiore Policlinico, 20122 Milan, Italy; luca.ruggiero@policlinico.mi.it
3. Department of Otolaryngology and Head and Neck Surgery, Fondazione IRCCS Ca' Granda Ospedale Maggiore Policlinico, 20122 Milan, Italy; pasquale.capaccio@unimi.it (P.C.); sara.torretta@unimi.it (S.T.)
4. Department of Biomedical Surgical Dental Science, University of Milan, 20122 Milan, Italy
5. Internal Medicine Department, Respiratory Unit and Adult Cystic Fibrosis Center, Fondazione IRCCS Ca' Granda Ospedale Maggiore Policlinico, 20122 Milan, Italy
6. Laboratory of Clinical Microbiology, Department of Biomedical Science for Health, University of Milan, 20122 Milan, Italy; lorenzo.drago@unimi.it
7. Department of Clinical Sciences and Community Health, University of Milan, 20122 Milan, Italy
* Correspondence: francesco.folino@unimi.it

Received: 19 July 2020; Accepted: 31 August 2020; Published: 2 September 2020

Abstract: Otitis media (OM) is one of the most common diseases occurring during childhood. Microbiological investigations concerning this topic have been primarily focused on the four classical otopathogens (*Streptococcus pneumoniae*, *Haemophilus influenzae*, *Moraxella catarrhalis* and *Streptococcus pyogenes*) mainly because most of the studies have been conducted with culture-dependent methods. In recent years, the introduction of culture-independent techniques has allowed high-throughput investigation of entire bacterial communities, leading to a better comprehension of the role of resident flora in health and disease. The upper respiratory tract (URT) is a region of major interest in otitis media pathogenesis, as it could serve as a source of pathogens for the middle ear (ME). Studies conducted with culture-independent methods in the URT and ME have provided novel insights on the pathogenesis of middle ear diseases through the identification of both possible new causative agents and of potential protective bacteria, showing that imbalances in bacterial communities could influence the natural history of otitis media in children. The aim of this review is to examine available evidence in microbiome research and otitis media in the pediatric age, with a focus on its different phenotypes: acute otitis media, otitis media with effusion and chronic suppurative otitis media.

Keywords: otitis media; microbiota; upper respiratory tract; adenoid; middle ear; microbiota axes

---

1. Introduction

The human microbiota consists of ecological communities of commensal, symbiotic and pathogenic microorganisms that colonize several body sites, as the gastrointestinal tract, respiratory system, oral cavity, skin and female reproductive system [1]. In past years, microbiological investigations have been predominantly conducted with culture-dependent methods, therefore many sites in the human body have been considered sterile until recently. However, the introduction of culture-independent techniques has allowed investigation of entire bacterial communities [2], leading to a better comprehension of the role of resident flora in health and disease. These microorganisms and their products play indeed a critical role in the regulation of many homeostatic processes, including immune response and inflammation [3] and defense against pathogens [4]. A diseased alteration in the composition of these bacterial communities, defined dysbiosis, can therefore lead to many pathological conditions, including infections [5].

Most of these studies have been conducted with a marker gene analysis based on a broad-range PCR, using primers that target a segment of the 16SrRNA gene, a highly conserved region contained in bacterial genomes. This method, combined with next-generation sequencing technologies, permits the simultaneous characterization of an entire community [6]. This approach allows a fast and cost-effective analysis that provides a low-resolution view of a microbial community. However, there are also some limitations that should be taken into account when interpreting data derived from these studies: it is not possible to determine whether taxa detected are alive or dead, active or inactive, thus there is limited functional information; it is susceptible to over amplification bias, especially with low biomass samples such as middle ear fluid; as a short segment of 16SrRNA gene is amplified and sequenced, taxonomic resolution is usually limited to family or genus level; there is great variability depending on technical aspects as region selection, amplicon size, sampling, storage, sequencing approach, and bioinformatic analysis. Full-gene 16S rRNA gene sequencing and metagenome and metatrascriptome analyses may overcome some of these limitations but are less adopted as they are relative expensive and complex to perform [7]. Middle ear infections and diseases are widespread in pediatric age. Acute otitis media (AOM) is the most common bacterial infection in childhood [8] and the leading cause of antibiotic prescription in pediatric patients [9]; similarly, otitis media with effusion (OME) is prevalent in the first years of life, as up to 80% of children suffer from one or more episodes by 10 years of age; however, it should be considered that the prevalence of OME varies across population and could be difficult to define accurately, as this condition is often asymptomatic [10].

The upper respiratory tract (URT) is a region of major interest in otitis media pathogenesis: According to the Pathogen Reservoir Hypothesis (PRH), the adenoid pad serves as a source of pathogens that can grow in this region and further spread to the respiratory system and middle ear, leading to infections and diseases [11–14].

The URT extends from the nostrils to the portion of the larynx above the vocal cords and harbors the highest bacterial density in the whole respiratory system [15]; however, these bacterial communities have been studied with more effort and from an ecological perspective only in recent years, after the introduction of culture-independent techniques [16].

Scientific interest has been focused on the comprehension of the characteristics of a healthy URT microbiota and the mechanism that guarantees its balance, as mounting evidence shows that resident bacteria are able to inhibit colonization and growth of otopathogens [14,17,18]. Those microorganisms that are essential in maintaining balance and function of a bacterial community are defined keystone species (see Table 1 for definitions of common terms used in microbiota analysis). In the URT, *Dolosigranulum* spp. and *Corynebacterium* spp. have been identified as potential keystone species, as they have been associated with respiratory health and exclusion of otopathogens in several studies [19–22].

Reconstitution of healthy microbial communities through administration of probiotics for the prevention of middle ear diseases in children is a topic of major clinical and scientific interest. Several trials have been conducted, but results lack consistency [23,24]. Deepening our knowledge on the physiological features of the URT microbiota and understanding how modifications in its balance relate to the pathogenesis of otitis media could be of remarkable importance in developing probiotic therapies. Furthermore, middle ear microbiota involvement in this field has been gaining interest in recent years, although less studies are available in comparison with URT microbiota, due to the different feasibility in collecting samples.

The aim of this review is to examine evidence available in microbiome research on otitis media in children. We will describe the most important factors that impact on microbiota development in the first years of life and that could influence the natural history of otitis media; then, we will focus on otitis media phenotypes and discuss evidence available on URT and middle ear microbiome in different diseases.

Table 1. Definitions of common terms used in microbiota investigations.

Microbiota	Ecological communities of commensal, symbiotic and pathogenic microorganisms that colonize several body sites, as the gastrointestinal tract, respiratory system, oral cavity, skin, and female reproductive system
Microbiome	Genetic material of the microorganisms of a community
Keystone Species	Microorganisms with a great impact on an ecological community, considered important in maintaining its organization and function
Biodiversity	Number of OTUs in a community and their relative abundance. It is determined by richness (how many OTUs in a sample?) and evenness (how equally distributed relative abundances are in a sample?)
Alpha-Diversity	Diversity within sample: how abundant OTUs are in relation to others in the same sample?
Beta-Diversity	Measure that compares different microbial communities
Operational Taxonomic Unit (OTU)	Cluster of related sequences (usually with 97% or more similarity) that represent a taxonomic unit of a microorganism

2. Methods

The research was conducted on the PubMed database, including all evidences available until April 2020. MeSH terms as "otitis media", "microbiota", "child", "child, preschool" and "infant" were used. More articles were included combining the keywords "microbiota" and "microbiome" with terms as "acute otitis media", "otitis media with effusion", "chronic otitis media", "adenoid", "adenotonsillar", "nasopharyngeal", "middle ear".

A total of 91 potentially relevant studies were identified through this search strategy. After title and abstract analysis, 51 studies were excluded as non-pertinent, according to the following criteria: disease different from OM; site of investigation different from URT or ME; adult population; studies conducted on animals were also excluded, as the main focus of this review was to discuss evidence available in children. A total of 40 remaining articles were then selected for more detailed assessment, and 14 investigations were further excluded in this phase (see Figure 1 for more details on methods).

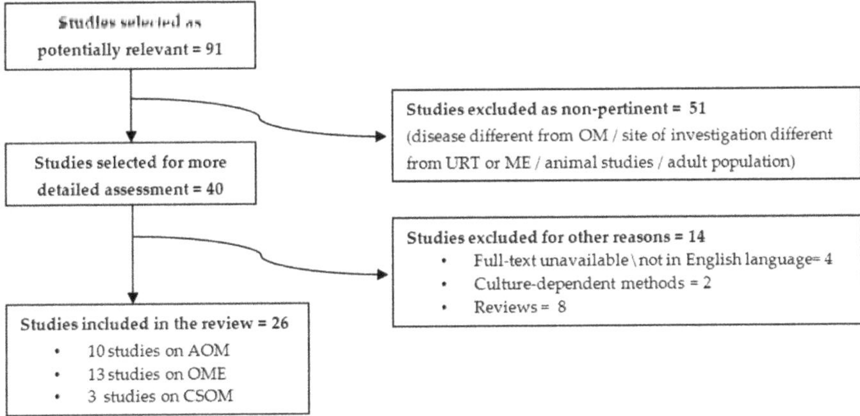

Figure 1. Search strategy conducted for this review. Legend: OM: Otitis media. AOM: Acute otitis media. OME: Otitis media with effusion. CSOM: Chronic suppurative otitis media. ME: Middle Ear. URT: Upper respiratory tract.

After this process, 26 studies were included in this review: 10 studies on acute otitis media (AOM, 1321 subject enrolled in all studies), 13 studies on otitis media with effusion (OME, 501 subjects enrolled in all studies), 3 studies on chronic suppurative otitis media (CSOM, 217 subjects enrolled in all studies).

3. Environmental Factors and Microbiota Development in the First Years of Life

The microbial communities that colonize the human organism are dynamic and change throughout life under the effect of several environmental factors, but infancy and early childhood represent the critical period in shaping their composition [25,26]. These external factors can impair the homeostatic functions mediated by the microbiota, leading to immediate consequences or impacting the health status in the later stages of life [27]. This is particularly evident for the URT microbiota, as this region is interconnected with middle ear, lower respiratory tract, and gastrointestinal tract, and represents the interface between these systems and the external environment.

Immediately after birth, in the first hours of life, the URT in healthy neonates becomes colonized by microorganisms of maternal origin [28]. Niche differentiation starts in the first week of life, with a predominance of *Staphylococcus* spp., followed by an enrichment of *Corynebacterium*, *Dolosigranulum*, and *Moraxella* [29].

The first months of life are of remarkable importance in the development of URT microbial communities and their composition: Biesbroek et al. described eight distinct microbiota profiles in the URT of healthy infants, showing that a distinct bacterial profile could be identified by the sixth week of life; moreover, this early bacterial colonization plays a pivotal role in the stability of microbial communities: profiles dominated by *Moraxella* and *Dolosigranulum/Corynebacterium* are associated with a stable microbiota and with lower rates of respiratory infections in later stages of life, while less stable profiles are associated with high abundance of *Haemophilus* and *Streptococcus* [30].

Theo et al. confirmed the role of *Corynebacterium* and found a positive role of *Alloiococcus* in the first year of life in the development of URT microbial communities; moreover, authors reported data on nasopharyngeal (NP) microbiota in children with respiratory diseases, concluding that some *Moraxella* spp. were associated with an increased risk of disease rather than respiratory health [31].

Several environmental factors, discussed below, can influence the shaping of the URT microbiota composition in the first years of life.

3.1. Delivery Route

As it is generally known, children born by caesarian-section (C-section) suffer from a higher incidence of respiratory illness and morbidity in comparison with children born by vaginal delivery [32,33].

In one of the first reports concerning nasopharyngeal microbiota and route of delivery, swabs from different body sites were collected from healthy neonates immediately after birth: Authors found that undifferentiated microbial communities in vaginally delivered children were similar to maternal vaginal microbiota, while those who were born by C-section had microbial communities resembling maternal skin surface [28].

A subsequent longitudinal study on this theme analyzed nasopharyngeal swabs collected from 102 children in the first 6 months of life, showing a predominance of bacteria previously associated to microbiome stability and respiratory health in early stages of life (*Moraxella*, *Corynebacterium*, and *Dolosigranulum*) in children born by vaginal delivery [29]. These microorganisms are likely derived from maternal skin (*Staphylococcus* and *Corynebacterium*) [34] or from vaginal tract (*Dolosigranulum*, *Staphylococcus*, or *Streptococcus*) [35].

However, by contrast, another study evidenced that differences related to delivery route are transient and disappear by six weeks of age, suggesting that the development of the microbiota in the postnatal period is more related to the body site that harbors a community [36].

3.2. Breastfeeding

Breastfeeding is a significant protective factor against infections [37,38]. This effect is related not only to the presence of antibacterial substances in maternal milk [39], as it is known that breastfeeding can significantly facilitate the development of a healthy microbiota.

Biesbroek et al. showed that breastfed infants develop a bacterial profile enriched by *Dolosigranulum* and *Corynebacterium* at six weeks of age in comparison with formula fed infants; moreover, *Dolosigranulum* abundance was inversely associated with wheezing episodes and a number of parental reported respiratory tract infections, even after correction for feeding type. [40].

Similar data were provided by Bosch et al.: Children who suffered from a higher number of respiratory infections had an aberrant nasopharyngeal microbiota development in the first month of life, that coincided with a prolonged reduction of *Dolosigranulum* and *Corynebacterium*; authors found that breastfeeding was an independent driver of this aberrant development, as a prolonged dominance of these bacteria was observed in breastfed infants. However, similarly to delivery route, these dissimilarities are transient and disappear around six months of age [41].

3.3. Antibiotic Therapy

Antibiotic therapy can significantly impair composition and balance of the microbiome [42]. This is particularly relevant in pediatric age, in which antibiotic prescription and misuse is quite common [43].

In the URT, antibiotic administration causes a reduction of the abundance of potential beneficial bacteria, such as *Dolosigranulum* and *Corynebacterium*, and an increase in *Haemophilus*, *Streptococcus*, and *Moraxella* [31]. Moreover, in children with AOM, a recent antibiotic therapy induces a reduction of *Streptococcaceae* and *Corynebacteriaceae* and an increased abundance of *Enterobacteriaceae* and *Pasturellaceae* in the URT [44]. Subsequent longitudinal studies confirmed how antibiotic treatment can induce a reduction in the abundance of potential beneficial bacteria, as *Dolosigranulum* and *Corynebacterium* [41,45].

3.4. Pneumococcal Vaccination

The introduction of the pneumococcal conjugate vaccination (PCV) in the pediatric population has led to an important reduction of OM episodes caused by the serotypes included in the vaccine [46]. On the other hand, the introduction of pneumococcal vaccination programs has resulted in important modifications in OM microbiology: *H. Influenzae* has become the most common otopathogen and OM episodes caused by *M. catarrhalis* have become more frequent; moreover, serotypes not included in PCVs have been more frequently identified as causative agents of diseases [47,48].

These findings suggest that the introduction of PCVs might have induced modifications in the composition of the microbial communities in the respiratory system. However, evidence supporting these findings has been focused prevalently on otopathogens, while studies conducted with high-throughput methods and looking at whole bacterial communities in the URT are lacking and show conflicting results.

In one of the first investigations on the whole NP microbiota and AOM, Hilty et al. reported that a previous exposure to PCV-7 in children with AOM was associated with reduced abundance of commensal families (*Streptococcaceae* and *Corynebacteriaceae*) [44].

A possible influence of this vaccination on NP microbiota was later reported by Biesbroek et al. In this investigation, NP swabs were collected from healthy children who received PCV-7 and from unvaccinated children: vaccination affected the URT microbiota causing a shift in composition and structure of the bacterial community, with an increase of *Veillonella*, *Prevotella*, *Fusobacterium*, *Leptotrichia*, *Actinomyces*, *Rothia*, and non pneumococcal streptococci, in addition to an increased bacterial diversity and inter-individual variability [49].

Longitudinal data on this theme were further provided in another study conducted in Switzerland by Mika et al., who compared NP microbiota in healthy children who were vaccinated with PCV-7 or PCV-13, showing that those who received PCV-13 had a more diverse and stable URT microbiota and a lower pneumococcal carriage rate compared to those who received PCV-7 [50].

However, in contrast to these findings, other available studies suggest that PCV might not have such a relevant impact on the URT microbiota. Faezel et al. performed a randomized controlled trial in Kenya comparing NP microbiota of children who received a 10-valent pneumococcal vaccine vs. children who received Hepatitis A vaccine. In this longitudinal study, NP swabs were collected before the administration of the vaccine and after 6 months. The authors found that PCV did not cause any significant alteration in the abundance or prevalence of otopathogens [51].

Moreover, a more detailed longitudinal study conducted in Gambia analyzed NP swabs collected periodically from birth to the first year of life. Children were divided in three groups according to vaccination schedule: Two groups received two different types of PCV-7, while the third group was composed by unvaccinated children. Again, bacterial communities were comparable across groups, as there were no significant differences in richness, diversity, and composition. Interestingly, PCV-7 vaccination reduced the nasopharyngeal carriage of vaccine serotypes, but pneumococcal carriage remained high among vaccinated infants, probably because of an immediate expansion of non-vaccine serotypes [52].

Interesting data were provided by Andrade et al. in a complex investigation that compared 53 children vaccinated with PCV-10 vs. 27 unvaccinated children. The strength of this study is the integrated metagenomic and transcriptomic analysis: no difference were found in nasopharyngeal carriage rates of *S. pneumoniae*, *S. aureus*, *H. influenzae*, or *M. catarrhalis* by either transcriptomic ormetagenomics analysis, but unvaccinated children had higher metabolic rates for *S. pneumoniae*, compared to PCV-10 vaccinated children [53].

Available evidence thus suggest that PCV has a direct impact on pneumococcal carriage, which in turn might indirectly affect the whole bacterial community in the URT. However, results are conflicting: a possible explanation could be found in the variation of pneumococcal carriage rates in relation to the geographic region and socio-economic status: the effects of PCV might indeed be different while considering developed or developing countries [54].

This particular theme was investigated in a study conducted in Fiji, in which NP microbiota from two ethnic groups (iTaukei and Fijians of Indian descent) was analyzed. These groups are known to have a different carriage prevalence of *S. pneumoniae* and a different burden of pneumococcal disease, which is higher in the iTaukei population. NP swabs were collected from 132 total children belonging to the two ethnic groups that were further divided in two subgroups based on whether children had been previously vaccinated or not with PCV-7. The vaccination had no overall impact on microbial diversity or composition, but significant modifications were evident when stratifying by ethnicity: vaccinated iTaukei children had a lower relative abundance of *Streptococcus* and *Haemophilus* compared with unvaccinated ones, while vaccinated Indian descent children had a higher relative abundance of *Dolosigranulum* compared with those unvaccinated [55].

3.5. Smoking

Studies conducted in adult subjects suggest that active smoking impairs URT microbiota composition [56]. It is likely that similar effects involve the pediatric population; however, studies concerning active and passive smoking effects on URT microbiome in children are lacking.

4. Acute Otitis Media

Acute otitis media (AOM) is defined by the presence of fluid in the middle ear associated to signs and symptoms of acute infection. It affects the majority of children in the first 3 years of life and becomes recurrent in almost 50% of cases [10]. Recurrent acute otitis media (RAOM) is defined as four or more AOM episodes in one year or three or more episodes in 6 months [57].

Laufer et al. performed one of the first studies comparing NP microbiome in children with AOM to healthy children. The authors evidenced that a higher relative abundance of *Corynebacterium* and *Dolosigranulum*, in addition to *Propionibacterium*, *Lactococcus*, and *Staphylococcus*, was associated with a lower incidence of pneumococcal colonization and AOM. The same study showed that a less diverse and a less even microbiota was associated with colonization by *S. pneumoniae*, highlighting the correlation between a higher biodiversity and better outcomes [20].

These data were confirmed in a subsequent investigation conducted by the same group on 240 children aged 6 months–3 years, that evidenced that a lower biodiversity was associated with a higher colonization rate not only from *S. pneumoniae* but also from *H. influenzae* and *M. catharralis*; moreover, authors compared diversity indices between health status and during an acute upper respiratory infection (URTI), showing that biodiversity was significantly higher in healthy children than during disease [19].

These findings on biodiversity during URTI are coherent with data provided by Hilty et al. that evidenced how NP bacterial density is lower in children during an AOM episode compared with the same in healthy status. Moreover, interesting insights were provided on how the infants' microbiota undergoes changes during an AOM episode, as the classical otopathogens predominated over commensal families (*Staphylococcaceae*, *Flavobacteriaceae*, *Carnobacteriaceae*, and *Comamonadaceae*) [44].

In 2017 Chonmaitree et al. performed a longitudinal study on 139 healthy neonates, followed since birth for the first 12 months of life or until the occurrence of the first AOM episode, collecting 971 swabs performed monthly and during an URTI or AOM. In particular, as it is known that URTI often precedes an AOM episode, authors studied the characteristics of the NP microbiome during transitional phase from URTI to AOM. Data revealed that an unstable microbiota during an URTI episode with the predominance of otopathogens were associated with the occurrence of symptomatic viral infection and with a higher risk of transition from URTI to AOM. Interestingly, otopathogens were not predominant during otherwise asymptomatic viral infections [45].

Evidence on otitis-prone children, i.e., those already suffering for RAOM were provided by Dirain et al.: Authors compared the microbial flora on adenoid tissue in a small group of subjects undergoing adenoidectomy for RAOM (*n* = 5) or obstructive sleep apnea (OSA) (*n* = 5), finding that the relative abundance of *S. pneumoniae* and *M. catharralis* was higher in the RAOM group [58].

A complex study with a higher sample size was subsequently performed on an Australian population, comparing NP microbiome of 103 healthy children vs. 93 otitis-prone children undergoing grommet insertion for RAOM, in order to identify potential protective genera. This investigation confirmed the pivotal role of *Dolosigranulum* and *Corynebacterium* in NP microbiome, as these two genera have been found to be significantly more abundant in the NP of healthy children compared with otitis-prone children. As for biodiversity, in contrast with previous findings, this study found that otitis-prone children had a significantly more diverse microbiome than controls. In addition, authors analyzed middle ear fluid (MEF) microbiome collected from children undergoing surgery from RAOM and performed a paired comparison with the NP microbiome of the same subject. Results showed that these two niches were not highly concordant: In particular, the interesting data is that *Alloiococcus* and *Turicella* have been found to be abundant in MEF but almost absent in the NP [22].

The MEF microbiome during an AOM episode was further investigated on 79 subjects aged 5–42 months. This report confirmed that the classical otopathogens are the predominant species in MEF during AOM: *S. pneumoniae* was dominant in 16% of samples, *H. influenzae* in 17%, and *M. catarrhalis* in 5.6%; moreover, *Turicell aotitidis* was detected as a clearly dominant bacteria in two samples, suggesting that it could be a rare but true causative agent; *Alloiococcus otitidis* was detected only in 3 samples; *Staphylococcus auricolaris* was predominant in two samples, but authors speculated that this finding could be related to potential contamination from the external auditory canal (EAC); however, *A. otitidis* and *T. otitidis* could be also related to EAC contamination [59].

Xu et al. compared the MEF microbiota during AOM episode to the NP microbiota analyzed on nasal wash (NW) samples: A significantly higher abundance of *A. otitidis* was detected in MEF during AOM, compared with NW in health and disease; authors concluded that the ME could harbor a resident microbiome that becomes different from NP after the onset of an infection. Moreover, NP microbiome was analyzed prior to the onset of AOM vs. at AOM onset: In line with previous data, NP microbiome during health was significantly more diverse than during AOM [60].

Paired analysis of NP and MEF microbiome during an AOM episode was subsequently performed on a larger population, collecting 286 NP swabs in children aged 0–6 years; 42/286 episodes were characterized by spontaneous tympanic membrane perforation (STMP), and thus, MEF microbiome was analyzed in these cases. Authors found that diversity was strictly related to age: in particular, older children had a higher richness and showed more personalized bacterial profiles, that develop toward the end of the sixth year of life. The transition to an adult-like microbiome appeared in children older than 3 years and was defined by an increase in *Staphylococcaceae* and *Corynebacteriaceae*. Furthermore, authors found concordance between NP and MEF microbiome when the predominant bacteria in MEF was *S. pyogenes*, *H. influenzae*, or *S. pneumoniae*. However, even this event appeared to be age-related, as the concordance between NP and MEF microbiome became weaker as children got older. Authors thus concluded that the NP microbiota does not necessarily resembles the one in ME: The URT in children with AOM serves as a moderate proxy for MEF at a very young age but becomes more diverse at a more advanced age [61].

The most frequently observed complication of AOM in clinical practice is the spontaneous tympanic membrane perforation (STMP) [62]. However, evidence on microbiota in children with history of RAOM with STMP is lacking. We believe that this condition represents a distinct phenotype of disease in otitis-prone children [63], and more effort should be directed to this category of patients, since their clinical management is often very challenging, and the most important AOM preventive measurements are often less effective [64–66].

Man et al. conducted a study on 94 children with tympanostomy tubes who suffered from ear discharge. In this case, authors observed a substantial concordance between paired NP and MEF microbiota, thus supporting the pathogen reservoir hypothesis: in particular, *Pseudomonas aeruginosa*, *Staphylococcus aureus*, *Streptococcus pyogenes*, *Turicella otitidis*, *Klebsiella pneumoniae*, and *Haemophilus* spp. were correlated between these two sites. *Moraxella* spp., *Streptococcus pneumoniae*, and *Corynebacterium/Dolosigranulum* were predominant in NP rather than in MEF, confirming their role as keystone bacteria of the URT; by contrast, *Turicella*, *P. aeruginosa* and *S. aureus* were strongly associated to MEF. Of interest, abundance of *Corynebacterium* and *Dolosigranulum* in NP related to a shorter course of the disease and better clinical outcomes [67].

Evidence available on AOM display that *Dolosigranulum* and *Corynebacterium* might act as potential keystone taxa in the URT, as they have been associated to a healthy status and to a lower colonization rate by otopathogens such as *S. pneumoniae*. Moreover, studies conducted on MEF identify *A. otitidis* and *T. otitidis* as possible novel otopathogens, although the theme of sample contamination from the EAC deserves major clarification.

An overview on microbiome study in AOM previously discussed is reported in Table 2.

Table 2. Overview of investigations on microbiota and acute otitis media discussed in this review.

Title (Year of Publication) [Ref]	Study Design	N. of Subjects	Age	Site of Investigation	Main Findings
Microbial Communities of the Upper Respiratory Tract and Otitis Media in Children (2011) [20]	Comparison of NP microbial communities in children with and without OM	108 (25 with AOM; 83 without AOM)	6–78 m	NP	• Microbial communities with *S. pneumoniae* were significantly less diverse and less even • Higher relative abundance of *Corynebacterium* and *Dolosigranulum*, in addition to *Propionibacterium*, *Lactococcus*, and *Staphylococcus*, was associated with a lower incidence of pneumococcal colonization and lower risk of AOM
Nasopharyngeal Microbiota in Infants with Acute Otitis Media (2012) [44]	Comparison of NP microbial communities in children with and without OM	163 (153 with AOM; 10 without AOM)	<2 y	NP	• NP bacterial density was lower during an AOM episode in comparison to health • Otopathogens predominated over commensal families during AOM
Upper Respiratory Tract Microbial Communities, Acute Otitis Media Pathogens, and Antibiotic Use in Healthy and Sick Children (2012) [19]	Comparison of NP microbial communities in healthy children vs. children with URTI with and without concurrent AOM	240 (73 healthy subjects; 95 subjects with URTI without concurrent AOM; 72 subjects with URTI with concurrent AOM)	6 m–3 y	NP	• Lower diversity was associated with a higher colonization rate by *S. pneumoniae*, *H. influenzae*, and *M. catarrhalis* • Biodiversity levels were significantly higher in healthy children than during disease • Children with antibiotic use in the past 6 months and a higher abundance of *Lactococcus* and *Propionibacterium* had a lower risk of AOM • Children with no antibiotic use in the past 6 months, a low abundance of *Streptococcus* and *Haemophilus*, and a high abundance of *Corynebacterium* and *Dolosigranulum* had a lower risk of AOM

Table 2. *Cont.*

Title (Year of Publication) [Ref]	Study Design	N. of Subjects	Age	Site of Investigation	Main Findings
Nasopharyngeal microbiota in infants and changes during viral upper respiratory tract infection and acute otitis media (2017) [45]	NP microbiota analysis of children followed from near birth for the first 12 months of life or until the occurrence of the first AOM episode. NP swabs collected monthly or during each URTI or AOM episode.	139 patients (971 samples)	<1 y	NP	• Bacterial diversity was lower in culture-samples positive for *S. pneumoniae* and *H. influenzae* compared to cultured-negative samples • Otopathogen colonization was related to higher incidence of URTI • Higher abundance of otopathogens and lower abundance of *Pseudomonas, Myroides, Yersinia,* and *Sphingomonas* during URTI and AOM • Higher otopathogen abundance during symptomatic viral infection but not during asymptomatic infection • An unstable microbiota during URTI and the predominance of otopathogens was associated with a higher risk of transition from URTI to AOM
The Adenoid Microbiome in Recurrent Acute Otitis Media and Obstructive Sleep Apnea (2017) [58]	Comparison of adenoid microbiota in subjects undergoing surgery for RAOM or OSA	10 (5 AOM; 5 OSA)	2–11 y	Adenoid	• *H. influenzae, M. catarrhalis, S. pneumoniae, P. aeruginosa,* and *S. aureus* were predominant in all samples • Relative abundance of *S. pneumoniae* and *M. catharralis* was higher in the RAOM group • The microbial profiles associated with RAOM were different from, but overlapped with OSA
Next-Generation Sequencing Combined with Specific PCR Assays To Determine the Bacterial 16S rRNA Gene Profiles of Middle Ear Fluid Collected from Children with Acute Otitis Media (2017) [59]	ME microbiota analysis during AOM episodes	79 subjects (90 samples)	5–42 m	ME	• *S. pneumoniae* was detected in 31% of samples, *H. influenzae* in 27%, *M. catarrhalis* in 20%, *Staphylococcus* spp. in 23%, *T. otitidis* in 5.6%, *A. otitidis* in 3.3% • *S. pneumoniae* was the dominant pathogen in 16% of samples, *H. influenzae* in 17%, *M. catarrhalis* in 5.6%

Table 2. *Cont.*

Title (Year of Publication) [Ref]	Study Design	N. of Subjects	Age	Site of Investigation	Main Findings
A microbiome case-control study of recurrent acute otitis media identified potentially protective bacterial genera (2018) [22]	Comparison of NP microbiota between children undergoing grommet insertion for RAOM (cases) vs. healthy children (controls); analysis of ME and EAC microbiota in cases	196 (93 cases; 103 controls)	<5 y	NP ME EAC	• Significantly higher abundance of *Corynebacterium* and *Dolosigranulum* was detected in NP of controls in comparison to cases • Paired NP and ME were not highly concordant: *Alloiococcus*, and *Turicella* were abundant in ME and EAC of cases and almost absent in NP of both groups • *Gemella* and *Neisseria* were typical of the NP in cases prevalent in the middle ear
Comparative Analysis of Microbiome in Nasopharynx and Middle Ear in Young Children with Acute Otitis Media (2019) [60]	Comparison of NP microbiota 1 to 3 weeks prior to onset of AOM vs. at onset of AOM; comparison of NP and ME microbiome during AOM	6	6–24 m	NP ME	• Significantly higher abundance of *A. otitidis* detected in MEF during AOM compared to NP in health and disease • NP microbiome during health had a significantly higher diversity than during AOM
Age-Dependent Dissimilarity of the Nasopharyngeal and Middle Ear Microbiota in Children with Acute Otitis Media (2019) [61]	NP microbiota analysis during AOM; Paired NP and ME microbiota analysis in children with STMP	286 (42/286 MEF from STMP)	0–6 y	NP ME	• Alpha and beta diversity levels were strictly related to age: older children had a higher richness and more personalized bacterial profiles • NP and MEF microbiome were concordant when MEF was dominated by *S. pyogenes*, *H. influenzae*, or *S. pneumoniae*
Respiratory Microbiota Predicts Clinical Disease Course of Acute Otorrhea in Children with Tympanostomy Tubes (2019) [67]	Paired analysis of NP and ME microbiota in children with otorrhea on tympanostomy tubes	94	<5 y	NP ME	• Microbiota composition of NP and ME differed significantly, although paired NP and ME samples were more similar than unpaired samples • *P. aeruginosa*, *S. aureus*, *S. pyogenes*, *T. otitidis*, *K. pneumoniae*, and *Haemophilus* spp. were correlated between NP and ME • *Moraxella* spp., *S. pneumoniae*, and *Corynebacterium/Dolosigranulum* were predominant in NP than in MEF • *Turicella*, *P. aeruginosa*, and *S. aureus* were strongly associated to ME • Higher abundance of *Corynebacterium* and *Dolosigranulum* in NP related to better clinical outcomes

AOM: Acute otitis media. OSA: Obstructive Sleep Apnea. RAOM: Recurrent Acute Otitis media. EAC: External Auditory Canal. NP: Nasopharynx. STMP: Spontaneous Tympanic Membrane Perforation. ME: Middle Ear. MEF: Middle Ear Fluid. URTI: Upper Respiratory Tract Infection.

5. Otitis Media with Effusion

Otitis media with effusion (OME) is defined as the presence of middle ear fluid without signs or symptoms of acute infection. It is defined chronic otitis media with effusion (COME) whether it persists for more than 3 months [10].

The first study on this topic with a high-throughput molecular approach was conducted by Liu et al., through the investigation of the microbiota of middle ear, adenoid, and tonsils in an 8-year old child with chronic middle ear effusion undergoing adenotonsillectomy and bilateral tympanic tube insertion. Middle ear microbiota was dominated by *Pseudomonadaceae*, and tonsil microbiota showed a predominance by *Streptococcaceae*; adenoid microbiota was the most complex, including *Pseudomonadaceae*, *Streptococcaceae*, *Fusobacteriaceae*, and *Pasteurellaceae*, and shared microorganisms found both in tonsils and middle ear, supporting the hypothesis that the adenoid pad could act as a reservoir for both of these sites [68].

Relevant new insights on OME were subsequently provided in an Australian study analyzing NP swabs, MEF, and adenoid specimens from 11 indigenous children undergoing surgery: MEF microbiome was characterized by low diversity indices and predominance of a single bacteria, in most cases *A. otitidis*, *H. influenzae*, or *Streptococcus* spp. In particular, *A. otitidis* was the most common taxa in MEF and was not detected in any NP or adenoid samples. Thus, authors speculated that its origin from NP was unlikely and that it could represent a typical microorganism of the ME niche; however, as *A. otitidis* is a known commensal of the ear canal [69], further studies were warranted to understand its role and the influence of the ear canal flora, especially in children who suffer from recurrent tympanic membrane perforations [70].

Fago-Olsen et al. analyzed microbiota of palatine tonsils and adenoids from children undergoing surgery for adenoid/tonsillar hyperplasia vs. subjects undergoing surgery for secretory otitis media (SOM), showing that several microorganisms were occasionally co-detected in both sites, but *H. influenzae*, *S. pneumoniae*, and *M. catarrhalis* were significantly more abundant in the adenoids and almost absent from palatine tonsils, indicating that adenoids but not palatine tonsils could act as main reservoir of pathogens leading to OM. However, it should be noted that this study did not include MEF microbiota analysis [71].

Data concerning dissimilarities between NP and MEF microbiome were provided in a following investigation including 10 children undergoing adenotonsillectomy and grommet insertion for OME. The authors reported that adenoid and tonsil microbiota shared a higher similarity than adenoid and ME, thus questioning the PRH in OME. According to previous findings, *Alloiococcus* and *Turicella* were detected only in MEF samples; however, the most abundant genera in middle ear were *Fusobacterium* and *Staphylococcus* [72]. These data were subsequently confirmed in an investigation by Ari et al. on a larger population of children with OME: ME microbiome was characterized by a significant predominance of *Alloicoccus otitidis* (44%), *Turicella otitidis* (6%), and *Staphylococcus auricularis* (3%), while adenoid harbored a high relative abundance of *Rothia*, *Staphylococcus*, and *Granulicatella*. As for diversity indices, no significant dissimilarities in alpha-diversity were found between MEF and adenoid niches [73].

The potential role of *A. otitidis* as a key bacteria of the ME was confirmed in an investigation by Chan et al., through the analysis of paired MEF samples and adenoid swabs from children undergoing grommet insertion for OME and of adenoid swabs from healthy subjects. Data evidenced a different composition in microbial communities between paired MEF and adenoid, as 13 of the 17 most abundant genera showed a statistically significant difference in relative abundance. In particular, *A. otitidis* was the predominant OTU in MEF (23% mean relative abundance), while it was almost absent in adenoid samples (<1% relative abundance). Interestingly, this taxa was found in greater abundance in children with unilateral effusion. Authors postulated that the dissimilarities between the MEF and adenoid microbiota could question the PRH in children with OME: Adenoidal hypertrophy and Eustachian tube dysfunction predispose to OME, but subsequent modifications in the ME environment determine an unbalance in the local flora with the predominance of a certain microorganism that can potentially lead to acute disease [74].

Similarly, caution when using nasopharyngeal microbiota as a proxy for ME was warranted by Boers et al. in an investigation comparing NP and ME microbiota in children with gastro-esophageal reflux (GER) associated OM vs. children who suffered from OM without GER. Authors enrolled 30 subjects with RAOM, COME or both undergoing tympanostomy tube placement, identifying *Alloiococcus* spp. and *Turicella* spp. as the most abundant taxa in MEF while absent in NP samples. As for GER, no apparent effects were found on the NP and ME microbiota in the two groups [75].

A more recent investigation conducted in a tertiary hospital in China analyzed ME and adenoid microbiota from children undergoing surgery for OME and adenoid hypertrophy (AH) vs. adenoid microbiota from subjects without ear disease undergoing adenotonsillectomy for OSA. ME was dominated by *Haemophilus* (14.75%), followed by *Staphylococcus* (9.37%) and *Halomonas* (7.85%); moreover, in contrast with previous findings, *Alloiococcus otitidis* had low relative abundance in this site (3.75%), and *Turicella* was not reported at all among the most abundant genera: Authors stated that these differences with previous findings could be attributable to variation in sampling methods, sample size or geographical location. Four taxa were found to be significantly differentially abundant between ME and adenoid in OME group (*Streptococcus*, *Neisseria*, *Alloprevotella*, and *Actinobacillus*), while the classical otopathogens were commonly found both in adenoid and ME in all OME patients. Adenoid microbiota in controls was composed predominantly by *Haemophilus* (15.96%), *Streptococcus* (13.33%), and *Moraxella* (12.28%); however, no significant differences in relative abundances of these genera were found in adenoids of OME patients vs. controls. According to this data and to previous findings, authors concluded that the dissimilarities in microbial compositions between these two niches challenge the PRH in OME [76].

The potential reservoirs for ME microbiome in children with OME were investigated by Chan et al.: MEF analysis showed similar results to the previous studies, as ME was dominated by *A. otitidis*, followed by *Haemophilus*, *Moraxella*, *Staphylococcus*, and *Streptococcus*; the EAC microbiome was mostly constituted by *A. otitidis*, *Staphylococcus* and *Pseudomonas* with rare otopathogens, whereas adenoid microbiome was composed prevalently by otopathogens, with rare EAC genera such as *Alloiococcus*. Basing on this data and on the previous study, authors concluded that both EAC and NP could act as a reservoir for the middle ear in children with OME. However, as bacterial translocation across an intact tympanum has not been demonstrated yet, a membrane perforation (spontaneous or iatrogenic) is probably needed to allow bacteria to translocate from EAC to ME. Unfortunately, a history of previous perforations in this cohort is not available [77].

Another pivotal genera in OME pathogenesis is *Haemophilus*, as highlighted in a study on ME microbiome in 55 children with chronic middle ear effusion: the most abundant genera were *Haemophilus* (relative abundance 22.54%), *Moraxella* (11.11%), *Turicella* (7.84%), *Alcaligenaceae* (5.84%), *Pseudomonas* (5.40%), and *Alloiococcus* (5.08%). Moreover, children were grouped by age, hearing loss, and mucin type expression in MEF: *Haemophilus* was significantly more abundant in children with hearing loss and was associated to MEF containing MUC5B and MUC5A, suggesting a correlation between hearing loss and mucin content in relationship to *Haemophilus* abundance [78].

Kolbe et al. provided data on 50 children undergoing tube placement for COME with a more detailed taxonomic resolution. In contrast to previous data that observed a predominance by *Alloiococcus*, *Moraxella*, or *Haemophilus* in MEF, in this study, microbial communities were highly variable, and the classical otopathogens were absent in about half of the samples. Moreover, authors compared subject based on whether they had a history of lower airway disease (asthma or bronchiolitis): *Haemophilus*, *Staphylococcus*, and *Moraxella* were significantly more abundant in children with lower airway diseases, while *Turicella* and *Alloiococcus* were less prevalent; in addition, ME microbial communities in children with history of asthma/bronchiolitis were significantly less diverse than children who had only COME [79].

Nasopharyngeal microbiome is less diverse in children suffering from OME than in controls, as highlighted by two case-control studies [80,81]. In particular, Walker et al. showed that the nasal microbiome in children with OME is composed of a higher abundance of pathogens, with a lower abundance of commensals as alpha-hemolytic Streptococci and *Lactococcus*. Moreover, cluster analysis revealed that profiles dominated by *Corynebacterium*, *Streptococcus*, or *Moraxella* were associated with COME, while healthy children had a more mixed bacterial profile with higher abundance of commensals [81].

In conclusion, investigations on OME discussed above confirm the role of the known otopathogens, in particular *H. influenzae*, as the predominant taxa in MEF during disease. Moreover, as previously described for AOM, *A. otitidis* and *T. otitidis* are frequently identified as abundant members of the ME microbiota. Studies have so far failed to define the possible reservoir for ME microbiome, and it is not possible to exclude a sample contamination from the EAC, especially in a low biomass environment as the ME. Concerning this theme, we believe that further studies should also be focused on patients with a history of tympanic membrane perforation, which might be the entryway for microorganisms that colonize the EAC.

An overview on microbiome study in OME previously discussed is reported in Table 3.

Table 3. Overview of investigations on microbiota and otitis media with effusion discussed in this review.

Title (Year of Publication) [Ref]	Study Design	N. of Subjects	Age	Site of Investigation	Main Findings
The Otologic Microbiome: A Study of the Bacterial Microbiota in a Pediatric Patient with Chronic Serous Otitis Media Using 16SrRNA Gene-Based Pyrosequencing (2011) [68]	Microbiota analysis in ME, adenoid, and tonsil specimens from one pediatric patient with chronic serous otitis media undergoing adenotonsillectomy and bilateral tympanic tube insertion	1	8 y	Adenoid ME Tonsil	• *Pseudomonadaceae* were predominant in ME • *Streptococcaceae* were predominant in tonsil • Adenoid microbiota included multiple predominant bacteria: *Pseudomonadaceae, Streptococcaceae, Fusobacteriaceae,* and *Pasteurellaceae* • Adenoid microbiota shared bacteria found both in tonsils and middle ear
The microbiome of otitis media with effusion in Indigenous Australian children (2015) [70]	MEF, NP, and adenoid microbiota analysis in children undergoing surgical treatment for OME	11	3–9 y	NP Adenoid ME	• ME microbiota was dominated by *A. otitidis, H. influenzae,* or *Streptococcus* spp. • *A. otitidis* was the most common OTU in MEF and was not detected in any NP or adenoid samples • *Streptococcus* spp., *H. influenzae,* and *M. catarrhalis* were common to all sample types
The Microbiome of Otitis Media with Effusion (2016) [74]	ME and adenoid microbiota analysis in children undergoing adenoidectomy with ventilation tube insertion for chronic OME. Adenoid microbiota analysis from healthy subjects enrolled as controls	33 (23 subjects with OME; 10 healthy controls)	OME group: 1–8 y Control group: 1–12 y	ME Adenoid	• ME microbiota was dominated by *A. otitidis* (23% mean relative abundance), *Haemophilus* (22%), *Moraxella* (5%), and *Streptococcus* (5%) • Different microbial composition between paired MEF and adenoid: 13/17 of the most abundant genera showed a statistically significant difference in relative abundance • *A. otitidis* was the predominant OTU in MEF (23% mean relative abundance), while it was almost absent in adenoid samples (<1% relative abundance)
The Relationship of the Middle Ear Effusion Microbiome to Secretory Mucin Production in Pediatric Patients with Chronic Otitis Media (2016) [78]	Microbiota analysis and mucin detection in MEF collected from children undergoing myringotomy with tympanostomy tube placement for chronic OME	55	3–176 m	ME	• The most abundant genera were *Haemophilus* (relative abundance 22.54%), *Moraxella* (11.11%), *Turicella* (7.84%), *Alcaligenaceae* (5.84%), *Pseudomonas* (5.40%), and *Alloiococcus* (5.08%) • *Haemophilus* was significantly more abundant in children with hearing loss and was associated to samples containing secretory mucins as MUC5B and MUC5A

Table 3. *Cont.*

Title (Year of Publication) [Ref]	Study Design	N. of Subjects	Age	Site of Investigation	Main Findings
Identification of the Bacterial Reservoirs for the Middle Ear Using Phylogenic Analysis (2017) [77]	ME and EAC microbiota analysis in children undergoing surgery for OME. Adenoid pad and ME microbiota analysis data were included from a previous study	18	1–14 y	ME EAC	• The MEF microbiota was dominated by *A. Otitidis* (37.5%), *Haemophilus* (14.4%), *Moraxella* (10.0%), *Staphylococcus* (8.2%), and *Streptococcus* (3.8%) • The EAC had a high abundance of *Alloiococcus* (58.0%), *Staphylococcus* (20.8%), and *Pseudomonas* (3.2) with rare otopathogens • The adenoid microbiota had a high abundance of otopathogens with rare EAC genera: *Alloiococcus* (0.1% vs. 28.9%), *Haemophilus* (25.2% vs. 18.2%), *Staphylococcus* (0.2% vs. 10.8%), *Streptococcus* (12.7% vs. 4.2%), and *Pseudomonas* (0 vs.2.1%, respectively)
Pathogen reservoir hypothesis investigated by analyses of the adenotonsillar and middle ear microbiota (2018) [72]	Adenoid, middle ear, and tonsil microbiota analysis in children undergoing surgical treatment for OME	10	5–10 y	Adenoid ME Tonsil	• The most abundant genera in all sites were *Fusobacterium, Haemophilus, Neisseria,* and *Porphyromonas* • Higher proportion of *Haemophilus* and *Moraxella* in the adenoid than ME • *Alloiococcus* and *Turicella* were detected only in MEF samples • Adenoid and tonsil microbiota shared a higher similarity than adenoid and ME
Characterization of the nasopharyngeal and middle ear microbiota in gastroesophageal reflux-prone versus gastroesophageal reflux non-prone children (2018) [75]	Analysis of NP and ME microbiota in children suffering from GER-associated OM vs. OM only undergoing surgical treatment for RAOM, COME, or both	30 (9 subjects with GER-associated OM; 21 subjects with OM without GER)	GER group 1.3–6 y No GER group 0.8–12–8 y	NP ME	• No effect of GER on NP and ME microbiota in the two groups • *Alloiococcus* spp. and *Turicella* spp. were the most common taxa in MEF and were not detected in any NP swab

Table 3. *Cont.*

Title (Year of Publication) [Ref]	Study Design	N. of Subjects	Age	Site of Investigation	Main Findings
The Adenoids but Not the Palatine Tonsils Serve as a Reservoir for Bacteria Associated with Secretory Otitis Media in Small Children (2019) [76]	Adenoid and tonsillar microbiota analysis in children undergoing surgical treatment for hyperplasia of adenoids/tonsils without infection (HP group) vs. children undergoing surgery for SOM	28 (112 samples) 14 subjects in HP group; 14 subjects in SOM group)	HP group 24–65 m SOM group 15–59 m	Adenoid Tonsils	• The number of OTUs detected in the adenoids from the HP group was significantly lower compared to the number detected in adenoids from SOM group • Streptococcus was the most abundant genus (average 25.6%) followed by *Fusobacterium* (11.1%) and *Haemophilus* (10.3%) • Microbial communities were significantly different between the adenoid and tonsil samples • *S. pneumoniae* was significantly more abundant in the adenoids of HP group compared to adenoids of SOM group • *Fusobacterium nucleatum* was abundant in the adenoids of HP group but was almost in the adenoids of SOM group • The classical otopathogens (*H. influenzae, S. pneumoniae,* and *M. catarrhalis*) were significantly more abundant in the adenoids than in the tonsils
Nasal microbial composition and chronic otitis media with effusion: A case-control study (2019) [81]	Comparison of nasal microbiota between children undergoing surgery for COME vs. healthy subjects	178 (73 cases; 105 controls)	Case group: mean age 47.5 m Control group: mean age 49.6 m	Nasal (anterior nares)	• Children with COME had lower diversity than healthy controls • Children with COME had a higher abundance of otopathogens and lower abundance of commensals as Haemolytic Streptococci and *Lactococcus* • Profiles that were *Corynebacterium-dominated* or *Moraxella-dominated* were associated with COME
Altered Middle Ear Microbiome in Children with Chronic Otitis Media with Effusion and Respiratory Illnesses (2019) [79]	Comparison of ME microbiota children with chronic OME and history of lower airways disease (asthma or bronchiolitis) vs. children with chronic OME without history of lower airways disease	50 (13 with history of lower airway disease)	3–176 m	ME	• The ME microbiome was significantly less diverse in children with lower airway disease • *Haemophilus, Staphylococcus,* and *Moraxella* were significantly more abundant in ME of children with lower airways disease
Analysis of the Microbiome in the Adenoids of Korean Children with Otitis Media with Effusion (2019) [80]	Adenoid microbiota comparison between children undergoing surgery for OME vs. children without undergoing surgery for obstructive symptoms	32 [16 subjects with OME; 16 subjects without OME)	19 m–15 y	Adenoid	• Diversity levels were lower in the OME group • *Haemophilus* was the most abundant in the OME group • *Prevotella, Delftia,* and *Corynebacterium* were the dominant genera in the OME group

Table 3. *Cont.*

Title (Year of Publication) [Ref]	Study Design	N. of Subjects	Age	Site of Investigation	Main Findings
The bacteriome of otitis media with effusion: does it originate from the adenoid? (2019) [73]	Adenoid and ME microbiota analysis in children undergoing surgery for OME	25	1.5–9 y	Adenoid ME	• ME microbiome was dominated by *A. otitis* (44%), *T. otitidis* (6%), and *S. auricularis* (3%) • Adenoid microbiome was dominated by *Rothia*, *Staphylococcus*, and *Granulicatella* • No statistically significant difference in alpha diversity between the two niches; adenoid samples clustered in the beta diversity graph
The microbiomes of adenoid and middle ear in children with otitis media with effusion and hypertrophy from a tertiary hospital in China (2020) [76]	Adenoid and ME microbiota analysis in children undergoing surgical treatment for OME vs. adenoid microbiota analysis in children undergoing surgery for OSA without ear disease	30 (15 in OME group; 15 in OSA group)	OME group 60–108 m OSA group 8–96 m	Adenoid ME	• ME in OME was dominated by *Haemophilus* (14.75%), *Staphylococcus* (9.37%), and *Halomonas* (7.85%) • Low abundance of *A. otitis* (3.75%) in ME in OME group • Adenoid microbiota in OME group was dominated by *Haemophilus* (21.87%), *Streptococcus* (19.65%), and *Neisseria* (5.8%) • Adenoid microbiota in OSA was dominated by *Haemophilus* (15.96%), *Streptococcus* (13.33%), and *Moraxella* (12.28%) • No significant differences in alpha-diversity between ME and adenoids in OME group • Beta diversity analyses showed that the microbiome structure of ME was dissimilar the adenoid one in OME patients: taxa found to be significantly differentially abundant between these two sites were *Streptococcus*, *Neisseria*, *Alloprevotella*, and *Actinobacillus*

OME: Otitis Media with Effusion. COME: Chronic Otitis Media with Effusion. NP: Nasopharynx. ME: Middle Ear. MEF: Middle Ear Fluid. GER: Gastro-esophageal reflux. RAOM: Recurrent acute otitis media. SOM: Secretive otitis media. EAC: External auditory Canal. OTU: Operational Taxonomic Unit. OSA: Obstructive sleep apnea.

6. Chronic Suppurative Otitis Media

Chronic suppurative otitis media (CSOM) is defined as a chronic inflammation of the middle ear and mastoid cavity, with recurrent or persistent ear discharge through a non-intact tympanic membrane [10]. Less evidence is available on microbial communities in pediatric patients suffering from this condition.

Neef et al. compared 24 children with CSOM undergoing mastoid surgery to 22 healthy controls undergoing ear surgery for other conditions as cochlear implantation or benign brain tumor removal. Microbiota analysis and conventional culture were performed on swabs collected from middle ear and mastoid cavity during surgery. Authors did not observe a typical bacterial profile associated to CSOM, but highlighted the limits of the conventional culture-based approach, as no bacteria were detected by culture in healthy subjects. By contrast, molecular analysis detected potential pathogens as *Staphylococcus*, *Pseudomonas*, and *Haemophilus* even in healthy controls. As for diversity, authors observed a major inter-personal difference among CSOM patients, whereas this finding was not observed for controls. This data supported the hypothesis that microbial communities' disruption and dysbiosis could be implicated in CSOM pathogenesis [82].

These dissimilarities among patients suffering from CSOM are age-related, as reported by Minami et al. In this investigation, middle ear swabs were collected during surgery in pediatric and adult patients undergoing tympanoplasty for wet or dry COM vs. subjects undergoing surgery from other conditions than otitis media. *Proteobacteria* was the predominant phylum detected in normal subjects, both adults and children. However, the normal middle ear microbiota differed significantly according to age: Authors concluded that this dissimilarity between adults and children could be related to the higher incidence of *Staphylococcus* (*Firmicutes* phylum) in adults. Subjects with active inflammation and wet COM had a lower abundance of *Proteobacteria* and a higher incidence of *Firmicutes*: Authors warranted this finding to be considered in the pathogenesis of active inflammation in COM, in relation to the potential penetration of several exogenous pathogens through a chronic perforation. On the other hand, microbiome of dry COM was not significantly different from normal middle ear [83].

Santos-Cortez et al. previously performed an investigation comparing ME and EAC microbiome in 16 indigenous Filipino subject with chronic tympanic membrane perforation, showing that the microbial communities between these two niches were similar, probably due to a cross-contamination process through the perforated eardrum. Moreover, authors investigated microbiota composition in subjects who were carrier of the A2ML1 gene, which encodes an alpha-2 macroglobulin-like 1 protein, previously identified as a genetically determined risk factor for of otitis media [84]. Authors detected a higher relative abundance of *Fusobacterium*, *Porphyromonas*, *Peptostreptococcus*, *Parvimonas*, and *Bacteroides* in the ME of A2ML1-carrier patients, while *Alloiococcus*, *Staphylococcus*, *Proteus*, and *Haemophilus* were more abundant in ME of non-carrier subjects. Authors speculated that the expected loss-of-function of A2ML1 protein could influence ME microbiota composition promoting survival and growth of specific microorganism. This findings warrant further investigations on the relationship between host genotype and microbiota in OM [85].

Evidence on CSOM is lacking and does not show peculiar features of microbial communities in this OM phenotype. Moreover, investigations discussed above include both adults and children, thus it is difficult to draw any general conclusion in the pediatric population. The penetration of microorganisms residing in the EAC from the chronic tympanic membrane perforation has been considered in the pathogenesis of the active inflammation in CSOM, but further studies are needed to define with major detail this aspect.

An overview on microbiome study in CSOM previously discussed is reported in Table 4.

Table 4. Overview of investigations on microbiota and chronic suppurative otitis media discussed in this review.

Title (Year of Publication) [Ref]	Study Design	N. of Subjects	Age	Site of Investigation	Main Findings
Molecular Microbiological Profile of Chronic Suppurative Otitis Media (2016) [82]	Comparison of ME and mastoid microbiota in patients with CSOM undergoing surgery vs. healthy controls	46 (24 subjects with CSOM; 22 healthy subjects)	6 m–85 y	ME Mastoid cavity	• No typical bacterial profile associated to CSOM • No bacteria were detected by culture in healthy subjects, while molecular analysis detected potential pathogens such as *Staphylococcus, Pseudomonas,* and *Haemophilus* • Inter-personal difference in diversity levels among CSOM patients but not among controls
Microbiomes of the Normal Middle Ear and Ears with Chronic Otitis Media (2017) [83]	ME microbiota analysis in patients undergoing tympanoplasty for wet or dry COM vs. subjects undergoing surgery from other conditions than otitis media	155 (67 healthy subjects; 44 subjects with COM without active infection; 44 subjects with COM with active infection)	1–84 y	ME	• The normal middle ear microbiota differed significantly according to age: in particular, a higher incidence of *Staphylococcus* (*Firmicutes* phylum) was detected in adults • Microbiome of dry COM was not significantly different from normal middle ear • Lower abundance of *Proteobacteria* and higher incidence of *Firmicutes* in subjects with active inflammation and wet COM
Middle ear microbiome differences in indigenous Filipinos with chronic otitis media due to a duplication in the A2ML1 gene (2016) [85]	ME and EAC microbiota analysis in indigenous Filipinos with chronic otitis media; comparison of microbial communities in subjects carriers of A2ML1 variant vs. non carrier subjects	16 (11 subjects carriers of A2ML1 variant)	4–24 y	ME EAC	• Microbial communities between ME and EAC were similar • Higher relative abundance of *Fusobacterium, Porphyromonas, Peptostreptococcus, Parvimonas,* and *Bacteroides* in the ME of A2ML1-carrier patients • Higher relative abundance of *Alloiococcus, Staphylococcus, Proteus,* and *Haemophilus* in ME of non-carrier subjects

CSOM: Chronic Suppurative Otitis Media. COM: Chronic Otitis Media. ME: Middle Ear. EAC: External auditory canal.

7. Probiotic Therapy

Prevention of OM in children represents one of the most difficult aspects in the clinical management of these patients.

Restoration of dysbiosis through administration of probiotic strains is a preventive strategy that has gained major clinical and scientific interest in recent years in several diseases, including otitis media.

Probiotics are defined as "live microorganisms that, when administered in adequate amounts, confer a health benefit on the host" [86]. The introduction of high-throughput sequencing methods has allowed the investigation of entire bacterial communities and the identification of microorganisms associated to health status in various conditions.

As previously discussed, evidence on microbiota in children suffering from OM suggest that *Corynebacterium* spp. and *Dolosigranulum pigrum* are potential keystone taxa in the URT; thus, major interest has been directed towards these two microorganisms and their potential use as probiotics.

A detailed discussion of evidence available on probiotic therapy in OM goes beyond the scope of this review, as it has been recently extensively reviewed elsewhere.

A recent review by van den Broek et al. described novel insights on probiotic therapy in OM [87]. Basing on Koch's postulates, authors introduced the "probiotic postulates" to define the ideal probiotic strain to be used in clinical practice: The microorganism can be found in high abundance in health status and decreased abundance during disease; the microorganism can be isolated from a healthy organism and grown in pure culture; the cultured organism should promote health when introduced into a diseased organism; it should be possible to re-isolate these microorganisms as identical to the original agent from the healthy host. According to available evidence and to this postulates, authors identified *Dolosigranulum* as a prime candidate for the development of probiotic therapy.

However, current knowledge is still not sufficient to define probiotic efficacy for preventing OM. A recent systematic review included 13 studies on this subject, concluding that available evidence on probiotics use for the prevention of AOM is limited; among the various formulations, possible benefit could derive from nasal administration [88].

The most important limitations in evidence on this topic are poor to moderate quality of the investigations and great heterogeneity in route of administration (oral vs. intranasal), probiotic strains included in formulations, duration of therapy, and outcome measures.

8. Conclusions

The introduction of the modern molecular techniques and the subsequent investigations on microbial communities in the human organisms have changed our conceptions of health and disease and our approach to infectious conditions.

It is indeed well known that health and disease status are not merely determined by the presence or the absence of a pathogen but depend on a complex balance established among pathogens, resident microbiota, and host immune response.

Investigations previously described in this review have provided novel insights on the pathogenesis of middle ear diseases and led to the identification of both possible new causative agents and of potential protective bacteria, showing that imbalances in bacterial communities of the URT and ME could influence the natural history of otitis media in children.

However, scientific data on this topic are often difficult to compare because of methodological differences in specimen collection and analysis, in the site of investigation, and in data reporting. Moreover, a lack of standard diagnostic criteria for OM across countries often influences the enrollment phase and contributes to increase the heterogeneity among populations under investigation.

Another element that complicates data interpretation and deserves standardization is the use of different databases during OTUs assignment. This is a relevant issue that should be taken into consideration for two main reasons: different databases might lead to heterogeneous results; some taxa could be misclassified with certain databases, as reported for *A. otitidis* and *T. otitidis* [89].

We believe that future investigation should be focused on the following aspects:

- Defining standard criteria of specimen collection, analysis, and data reporting, in order to facilitate data comparison across studies;
- Deepening our knowledge on the impact of various exogenous factors that have been less explored, such as active/passive smoking, vaccines, and viral infections;
- Confirming the role of *Corynebacterium* and/or *Dolosigranulum* as keystone taxa, in order to evaluate their possible use as probiotics;
- Understanding the development of URT and ME microbiota at different ages, in order to identify a potential "window of opportunity" in which therapeutic interventions as probiotic administration could be more effective, before the establishment of a stable microbial community that could be modulated with difficulty;
- Investigating the concordance between NP and ME microbiota, in order to better define the role of adenoid pad as a proxy for ME;
- Providing data on microbial communities in ME, which is no longer considered a sterile site;
- Defining with major detail the features of NP and ME microbial communities in different OM phenotypes, in particular in children with recurrent STMP.

Author Contributions: P.M. and S.T. conceived the paper; F.F. conducted the literature search and wrote the manuscript; I.C. contributed to the literature search; P.M., S.T., L.D., P.C. and S.A. revised the paper; L.R. revised the technical aspects. All authors have read and agreed to the published version of the manuscript.

Funding: This research received no external funding.

Conflicts of Interest: The authors declare no conflict of interest.

References

1. The NIH HMP Working Group; Peterson, J.; Garges, S.; Giovanni, M.; McInnes, P.; Wang, L.; Schloss, J.A.; Bonazzi, V.; McEwen, J.E.; Wetterstrand, K.A.; et al. The NIH Human Microbiome Project. *Genome Res.* **2009**, *19*, 2317–2323. [CrossRef] [PubMed]
2. Patel, A.H.; Harris, K.A.; Fitzgerald, F. What is broad-range 16S rDNA PCR? *Arch. Dis. Child. Educ. Pract. Ed.* **2017**, *102*, 261–264. [CrossRef] [PubMed]
3. Samuelson, D.R.; Welsh, D.A.; Shellito, J.E. Regulation of lung immunity and host defense by the intestinal microbiota. *Front. Microbiol.* **2015**, *6*, 1085. [CrossRef] [PubMed]
4. Buffie, C.G.; Pamer, E.G. Microbiota-mediated colonization resistance against intestinal pathogens. *Nat. Rev. Immunol.* **2013**, *13*, 790–801. [CrossRef] [PubMed]
5. Honda, K.; Littman, D.R. The microbiome in infectious disease and inflammation. *Annu. Rev. Immunol.* **2012**, *30*, 759–795. [CrossRef] [PubMed]
6. Schuster, S.C. Next-generation sequencing transforms today's biology. *Nat. Methods* **2007**, *5*, 16–18. [CrossRef] [PubMed]
7. Knight, R.; Vrbanac, A.; Taylor, B.C.; Aksenov, A.; Callewaert, C.; Debelius, J.; González, A.; Kosciolek, T.; McCall, L.-I.; McDonald, D.; et al. Best practices for analysing microbiomes. *Nat. Rev. Genet.* **2018**, *16*, 410–422. [CrossRef]
8. Paradise, J.L.; Rockette, H.E.; Colborn, D.K.; Bernard, B.S.; Smith, C.G.; Kurs-Lasky, M.; Janosky, J.E. Otitis Media in 2253 Pittsburgh-Area Infants: Prevalence and Risk Factors During the First Two Years of Life. *Pediatrics* **1997**, *99*, 318–333. [CrossRef]
9. Marom, T.; Marchisio, P.G.; Tamir, S.O.; Torretta, S.; Gavriel, H.; Esposito, S. Complementary and Alternative Medicine Treatment Options for Otitis Media. *Medicine* **2016**, *95*, e2695. [CrossRef]
10. Schilder, A.G.; Chonmaitree, T.; Cripps, A.W.; Rosenfeld, R.M.; Casselbrant, M.L.; Haggard, M.P.; Venekamp, R.P. Otitis media. *Nat. Rev. Dis. Prim.* **2016**, *2*, 16063. [CrossRef]
11. Bernstein, J.M.; Reddy, M.S.; Scannapieco, F.A.; Faden, H.S.; Ballow, M. The Microbial Ecology and Immunology of the Adenoid: Implications for Otitis Media. *Ann. N.Y. Acad. Sci.* **1997**, *830*, 19–31. [CrossRef] [PubMed]
12. Nistico, L.; Kreft, R.; Gieseke, A.; Coticchia, J.M.; Burrows, A.; Khampang, P.; Liu, Y.; Kerschner, J.E.; Post, J.C.; Lonergan, S.; et al. Adenoid Reservoir for Pathogenic Biofilm Bacteria. *J. Clin. Microbiol.* **2011**, *49*, 1411–1420. [CrossRef] [PubMed]

13. Hoa, M.; Tomovic, S.; Nistico, L.; Hall-Stoodley, L.; Stoodley, P.; Sachdeva, L.; Berk, R.; Coticchia, J.M. Identification of adenoid biofilms with middle ear pathogens in otitis-prone children utilizing SEM and FISH. *Int. J. Pediatr. Otorhinolaryngol.* **2009**, *73*, 1242–1248. [CrossRef] [PubMed]

14. Bogaert, D.; De Groot, R.; Hermans, P. Streptococcus pneumoniae colonisation: The key to pneumococcal disease. *Lancet Infect. Dis.* **2004**, *4*, 144–154. [CrossRef]

15. Man, W.H.; Piters, W.A.A.D.S.; Bogaert, D. The microbiota of the respiratory tract: Gatekeeper to respiratory health. *Nat. Rev. Genet.* **2017**, *15*, 259–270. [CrossRef] [PubMed]

16. Piters, W.A.A.D.S.; Sanders, E.A.M.; Bogaert, D. The role of the local microbial ecosystem in respiratory health and disease. *Philos. Trans. R. Soc. B Biol. Sci.* **2015**, *370*, 20140294. [CrossRef]

17. Tano, K.; Grahn-Håkansson, E.; Holm, S.E.; Hellström, S. Inhibition of OM pathogens by alpha-hemolytic streptococci from healthy children, children with SOM and children with rAOM. *Int. J. Pediatr. Otorhinolaryngol.* **2000**, *56*, 185–190. [CrossRef]

18. Tano, K.; Olofsson, C.; Grahn-Håkansson, E.; Holm, S.E. In vitro inhibition of S. pneumoniae, nontypable H. influenzae and M. catharralis by alpha-hemolytic streptococci from healthy children. *Int. J. Pediatr. Otorhinolaryngol.* **1999**, *47*, 49–56. [CrossRef]

19. Pettigrew, M.M.; Laufer, A.S.; Gent, J.F.; Kong, Y.; Fennie, K.; Metlay, J.P. Upper Respiratory Tract Microbial Communities, Acute Otitis Media Pathogens, and Antibiotic Use in Healthy and Sick Children. *Appl. Environ. Microbiol.* **2012**, *78*, 6262–6270. [CrossRef]

20. Laufer, A.S.; Metlay, J.P.; Gent, J.F.; Fennie, K.; Kong, Y.; Pettigrew, M.M. Microbial Communities of the Upper Respiratory Tract and Otitis Media in Children. *mBio* **2011**, *2*, e00245–e00310. [CrossRef]

21. Bomar, L.; Brugger, S.D.; Yost, B.H.; Davies, S.S.; Lemon, K.P. Corynebacterium accolensReleases Antipneumococcal Free Fatty Acids from Human Nostril and Skin Surface Triacylglycerols. *mBio* **2016**, *7*, 01725. [CrossRef] [PubMed]

22. Lappan, R.; Imbrogno, K.; Sikazwe, C.; Anderson, D.; Mok, D.; Coates, H.; Vijayasekaran, S.; Bumbak, P.; Blyth, C.; Jamieson, S.E.; et al. A microbiome case-control study of recurrent acute otitis media identified potentially protective bacterial genera. *BMC Microbiol.* **2018**, *18*, 13. [CrossRef] [PubMed]

23. Tano, K.; Håkansson, E.G.; Holm, S.E.; Hellström, S. A nasal spray with alpha-haemolytic streptococci as long term prophylaxis against recurrent otitis media. *Int. J. Pediatr. Otorhinolaryngol.* **2002**, *62*, 17–23. [CrossRef]

24. Marchisio, P.G.; Santagati, M.C.; Scillato, M.; Baggi, E.; Fattizzo, M.; Rosazza, C.; Stefani, S.; Esposito, S.; Principi, N. Streptococcus salivarius 24SMB administered by nasal spray for the prevention of acute otitis media in otitis-prone children. *Eur. J. Clin. Microbiol. Infect. Dis.* **2015**, *34*, 2377–2383. [CrossRef] [PubMed]

25. Costello, E.K.; Lauber, C.L.; Hamady, M.; Fierer, N.; Gordon, J.I.; Knight, R. Bacterial Community Variation in Human Body Habitats Across Space and Time. *Science* **2009**, *326*, 1694–1697. [CrossRef]

26. Thomas, S.; Izard, J.; Walsh, E.; Batich, K.; Chongsathidkiet, P.; Clarke, G.; Sela, D.A.; Muller, A.J.; Mullin, J.M.; Albert, K.; et al. The Host Microbiome Regulates and Maintains Human Health: A Primer and Perspective for Non-Microbiologists. *Cancer Res.* **2017**, *77*, 1783–1812. [CrossRef]

27. Charbonneau, M.R.; Blanton, L.V.; DiGiulio, D.B.; Relman, D.A.; Lebrilla, C.B.; Mills, D.A.; Gordon, J.I. A microbial perspective of human developmental biology. *Nature* **2016**, *535*, 48–55. [CrossRef]

28. Dominguez-Bello, M.G.; Costello, E.K.; Contreras, M.; Magris, M.; Hidalgo, G.; Fierer, N.; Knight, R. Delivery mode shapes the acquisition and structure of the initial microbiota across multiple body habitats in newborns. *Proc. Natl. Acad. Sci. USA* **2010**, *107*, 11971–11975. [CrossRef]

29. Bosch, A.A.; Levin, E.; Van Houten, M.A.; Hasrat, R.; Kalkman, G.; Biesbroek, G.; Piters, W.A.A.D.S.; De Groot, P.-K.C.; Pernet, P.; Keijser, B.J.; et al. Development of Upper Respiratory Tract Microbiota in Infancy is Affected by Mode of Delivery. *EBioMedicine* **2016**, *9*, 336–345. [CrossRef]

30. Biesbroek, G.; Tsivtsivadze, E.; Sanders, E.A.M.; Montijn, R.; Veenhoven, R.H.; Keijser, B.J.F.; Bogaert, D. Early Respiratory Microbiota Composition Determines Bacterial Succession Patterns and Respiratory Health in Children. *Am. J. Respir. Crit. Care Med.* **2014**, *190*, 1283–1292. [CrossRef]

31. Teo, S.M.; Mok, D.; Pham, K.; Kusel, M.; Serralha, M.; Troy, N.; Holt, B.J.; Hales, B.J.; Walker, M.L.; Hollams, E.; et al. The infant nasopharyngeal microbiome impacts severity of lower respiratory infection and risk of asthma development. *Cell Host Microbe* **2015**, *17*, 704–715. [CrossRef] [PubMed]

32. Guibas, G.V.; Moschonis, G.; Xepapadaki, P.; Roumpedaki, E.; Androutsos, O.; Manios, Y.; Papadopoulos, N.G. Conception viain vitrofertilization and delivery by Caesarean section are associated with paediatric asthma incidence. *Clin. Exp. Allergy* **2013**, *43*, 1058–1066. [CrossRef] [PubMed]

33. Kristensen, K.; Fisker, N.; Haerskjold, A.; Ravn, H.; Simões, E.A.F.; Stensballe, L. Caesarean Section and Hospitalization for Respiratory Syncytial Virus Infection: A population-based study. *Pediatr. Infect. Dis. J.* **2015**, *34*, 145–148. [CrossRef] [PubMed]

34. Grice, E.A.; Segre, J.A. The skin microbiome. *Nat. Rev. Genet.* **2011**, *9*, 244–253. [CrossRef]

35. Mendling, W. Vaginal Microbiota. *Adv. Exp. Med. Biol.* **2016**, *902*, 83–93. [CrossRef]

36. Chu, D.M.; Ma, J.; Prince, A.L.; Antony, K.M.; Seferovic, M.D.; Aagaard, K.M. Maturation of the infant microbiome community structure and function across multiple body sites and in relation to mode of delivery. *Nat. Med.* **2017**, *23*, 314–326. [CrossRef]

37. Tarrant, M.; Kwok, M.K.; Lam, T.-H.; Leung, G.M.; Schooling, C.M. Breast-feeding and Childhood Hospitalizations for Infections. *Epidemiology* **2010**, *21*, 847–854. [CrossRef]

38. Duijts, L.; Jaddoe, V.W.V.; Hofman, A.; Moll, H.A. Prolonged and Exclusive Breastfeeding Reduces the Risk of Infectious Diseases in Infancy. *Pediatrics* **2010**, *126*, e18–e25. [CrossRef]

39. Labbok, M.; Clark, D.; Goldman, A.S. Breastfeeding: Maintaining an irreplaceable immunological resource. *Nat. Rev. Immunol.* **2004**, *4*, 565–572. [CrossRef]

40. Biesbroek, G.; Bosch, A.A.; Wang, X.; Keijser, B.J.F.; Veenhoven, R.H.; Sanders, E.A.; Bogaert, D. The Impact of Breastfeeding on Nasopharyngeal Microbial Communities in Infants. *Am. J. Respir. Crit. Care Med.* **2014**, *190*, 298–308. [CrossRef]

41. Bosch, A.A.T.M.; Wouter, A.A.d.S.P.; Van Houten, M.A.; Chu, M.L.J.N.; Biesbroek, G.; Kool, J.; Pernet, P.; De Groot, P.-K.C.M.; Eijkemans, M.J.C.; Keijser, B.J.F.; et al. Maturation of the Infant Respiratory Microbiota, Environmental Drivers, and Health Consequences. A Prospective Cohort Study. *Am. J. Respir. Crit. Care Med.* **2017**, *196*, 1582–1590. [CrossRef] [PubMed]

42. Sullivan, Å.; Edlund, C.; Nord, C.E. Effect of antimicrobial agents on the ecological balance of human microflora. *Lancet Infect. Dis.* **2001**, *1*, 101–114. [CrossRef]

43. Hicks, L.A.; Taylor, T.H.; Hunkler, R.J. U.S. Outpatient Antibiotic Prescribing, 2010. *N. Engl. J. Med.* **2013**, *368*, 1461–1462. [CrossRef] [PubMed]

44. Hilty, M.; Qi, W.; Brugger, S.D.; Frei, L.; Agyeman, P.; Frey, P.M.; Aebi, S.; Mühlemann, K. Nasopharyngeal Microbiota in Infants with Acute Otitis Media. *J. Infect. Dis.* **2012**, *205*, 1048–1055. [CrossRef]

45. Chonmaitree, T.; Jennings, K.; Golovko, G.; Khanipov, K.; Pimenova, M.; Patel, J.A.; McCormick, D.P.; Loeffelholz, M.J.; Fofanov, Y. Nasopharyngeal microbiota in infants and changes during viral upper respiratory tract infection and acute otitis media. *PLoS ONE* **2017**, *12*, e0180630. [CrossRef]

46. Pettigrew, M.M.; Alderson, M.R.; Bakaletz, L.O.; Barenkamp, S.J.; Hakansson, A.P.; Mason, K.M.; Nokso-Koivisto, J.; Patel, J.; Pelton, S.I.; Murphy, T.F. Panel 6: Vaccines. *Otolaryngol. Neck Surg.* **2017**, *156* (Suppl. 4), S76–S87. [CrossRef]

47. Ngo, C.C.; Massa, H.M.; Thornton, R.B.; Cripps, A.W. Predominant Bacteria Detected from the Middle Ear Fluid of Children Experiencing Otitis Media: A Systematic Review. *PLoS ONE* **2016**, *11*, e0150949. [CrossRef]

48. Gladstone, R.A.; Jefferies, J.M.; Tocheva, A.S.; Beard, K.R.; Garley, D.; Chong, W.W.; Bentley, S.D.; Faust, S.N.; Clarke, S.C. Five winters of pneumococcal serotype replacement in UK carriage following PCV introduction. *Vaccine* **2015**, *33*, 2015–2021. [CrossRef]

49. Biesbroek, G.; Wang, X.; Keijser, B.J.; Eijkemans, R.M.; Trzcinski, K.; Rots, N.Y.; Veenhoven, R.H.; Sanders, E.A.; Bogaert, D. Seven-Valent Pneumococcal Conjugate Vaccine and Nasopharyngeal Microbiota in Healthy Children. *Emerg. Infect. Dis.* **2014**, *20*, 201–210. [CrossRef]

50. Mika, M.; Maurer, J.; Korten, I.; Allemann, A.; Aebi, S.; Brugger, S.D.; Qi, W.; Frey, U.; Latzin, P.; Hilty, M. Influence of the pneumococcal conjugate vaccines on the temporal variation of pneumococcal carriage and the nasal microbiota in healthy infants: A longitudinal analysis of a case–control study. *Microbiome* **2017**, *5*, 85. [CrossRef]

51. Feazel, L.M.; Santorico, S.A.; Robertson, C.E.; Bashraheil, M.; Scott, J.A.G.; Frank, D.; Hammitt, L.L. Effects of Vaccination with 10-Valent Pneumococcal Non-Typeable Haemophilus influenza Protein D Conjugate Vaccine (PHiD-CV) on the Nasopharyngeal Microbiome of Kenyan Toddlers. *PLoS ONE* **2015**, *10*, e0128064. [CrossRef] [PubMed]

52. Kwambana-Adams, B.A.; Hanson, B.; Worwui, A.; Agbla, S.; Foster-Nyarko, E.; Ceesay, F.; Ebruke, C.; Egere, U.; Zhou, Y.; Ndukum, M.; et al. Rapid replacement by non-vaccine pneumococcal serotypes may mitigate the impact of the pneumococcal conjugate vaccine on nasopharyngeal bacterial ecology. *Sci. Rep.* **2017**, *7*, 1–11. [CrossRef] [PubMed]

53. Andrade, D.C.; Borges, I.C.; Bouzas, M.L.; Oliveira, J.R.; Fukutani, K.F.; Queiroz, A.T.; De Oliveira, C.I.; Barral, A.; Van Weyenbergh, J.; Nascimento-Carvalho, C. 10-valent pneumococcal conjugate vaccine (PCV10) decreases metabolic activity but not nasopharyngeal carriage of Streptococcus pneumoniae and Haemophilus influenzae. *Vaccine* **2017**, *35*, 4105–4111. [CrossRef]

54. Adegbola, R.A.; DeAntonio, R.; Hill, P.C.; Roca, A.; Usuf, E.; Hoet, B.; Greenwood, B.M. Carriage of Streptococcus pneumoniae and Other Respiratory Bacterial Pathogens in Low and Lower-Middle Income Countries: A Systematic Review and Meta-Analysis. *PLoS ONE* **2014**, *9*, e103293. [CrossRef] [PubMed]

55. Boelsen, L.K.; Dunne, E.M.; Mika, M.; Eggers, S.; Nguyen, C.D.; Ratu, F.T.; Russell, F.M.; Mulholland, E.K.; Hilty, M.; Satzke, C. The association between pneumococcal vaccination, ethnicity, and the nasopharyngeal microbiota of children in Fiji. *Microbiome* **2019**, *7*, 106. [CrossRef]

56. Charlson, E.S.; Chen, J.; Custers-Allen, R.; Bittinger, K.; Li, H.; Sinha, R.; Hwang, J.; Bushman, F.D.; Collman, R.G. Disordered Microbial Communities in the Upper Respiratory Tract of Cigarette Smokers. *PLoS ONE* **2010**, *5*, e15216. [CrossRef]

57. Teele, D.W.; Klein, J.O.; Rosner, B. Epidemiology of Otitis Media During the First Seven Years of Life in Children in Greater Boston: A Prospective, Cohort Study. *J. Infect. Dis.* **1989**, *160*, 83–94. [CrossRef]

58. Dirain, C.O.; Silva, R.C.; Collins, W.O.; Antonelli, P.J. The Adenoid Microbiome in Recurrent Acute Otitis Media and Obstructive Sleep Apnea. *J. Int. Adv. Otol.* **2017**, *13*, 333–339. [CrossRef]

59. Sillanpää, S.; Kramna, L.; Oikarinen, S.; Sipilä, M.; Rautiainen, M.; Aittoniemi, J.; Laranne, J.; Hyöty, H.; Cinek, O. Next-Generation Sequencing Combined with Specific PCR Assays To Determine the Bacterial 16S rRNA Gene Profiles of Middle Ear Fluid Collected from Children with Acute Otitis Media. *mSphere* **2017**, *2*, 00006–00017. [CrossRef]

60. Xu, Q.; Gill, S.; Xu, L.; Gonzalez, E.; Pichichero, M.E. Comparative Analysis of Microbiome in Nasopharynx and Middle Ear in Young Children With Acute Otitis Media. *Front. Genet.* **2019**, *10*, 1–7. [CrossRef]

61. Brugger, S.D.; Kraemer, J.G.; Qi, W.; Bomar, L.; Oppliger, A.; Hilty, M. Age-Dependent Dissimilarity of the Nasopharyngeal and Middle Ear Microbiota in Children With Acute Otitis Media. *Front. Genet.* **2019**, *10*, 555. [CrossRef] [PubMed]

62. Berger, G. Nature of spontaneous tympanic membrane perforation in acute otitis media in children. *J. Laryngol. Otol.* **1989**, *103*, 1150–1153. [CrossRef] [PubMed]

63. Torretta, S.; Marchisio, P. Otitis media in children: A proposal for a new nosological classification. *Int. J. Pediatr. Otorhinolaryngol.* **2017**, *93*, 174–175. [CrossRef] [PubMed]

64. Marchisio, P.G.; Nazzari, E.; Torretta, S.; Esposito, S.; Principi, N. Medical prevention of recurrent acute otitis media: An updated overview. *Expert Rev. Anti Infect. Ther.* **2014**, *12*, 611–620. [CrossRef]

65. Marchisio, P.G.; Esposito, S.; Bianchini, S.; Dusi, E.; Fusi, M.; Nazzari, E.; Picchi, R.; Galeone, C.; Principi, N. Efficacy of Injectable Trivalent Virosomal-Adjuvanted Inactivated Influenza Vaccine in Preventing Acute Otitis Media in Children With Recurrent Complicated or Noncomplicated Acute Otitis Media. *Pediatr. Infect. Dis. J.* **2009**, *28*, 855–859. [CrossRef]

66. Marchisio, P.G.; Consonni, D.; Baggi, E.; Zampiero, A.; Bianchini, S.; Terranova, L.; Tirelli, S.; Esposito, S.; Principi, N. Vitamin D Supplementation Reduces the Risk of Acute Otitis Media in Otitis-prone Children. *Pediatr. Infect. Dis. J.* **2013**, *32*, 1055–1060. [CrossRef]

67. Man, W.H.; Van Dongen, T.M.; Venekamp, R.P.; Pluimakers, V.G.; Chu, M.L.J.; Van Houten, M.A.; Sanders, E.A.; Schilder, A.G.M.; Bogaert, D. Respiratory Microbiota Predicts Clinical Disease Course of Acute Otorrhea in Children With Tympanostomy Tubes. *Pediatr. Infect. Dis. J.* **2019**, *38*, e116–e125. [CrossRef]

68. Liu, C.M.; Cosetti, M.K.; Aziz, M.; Buchhagen, J.L.; Contente-Cuomo, T.L.; Price, L.B.; Keim, P.; Lalwani, A.K. The Otologic MicrobiomeA Study of the Bacterial Microbiota in a Pediatric Patient With Chronic Serous Otitis Media Using 16SrRNA Gene-Based Pyrosequencing. *Arch. Otolaryngol. Head Neck Surg.* **2011**, *137*, 664–668. [CrossRef]

69. Frank, D.N.; Spiegelman, G.B.; Davis, W.; Wagner, E.; Lyons, E.; Pace, N.R. Culture-independent molecular analysis of microbial constituents of the healthy human outer ear. *J. Clin. Microbiol.* **2003**, *41*, 295–303. [CrossRef]

70. Jervis-Bardy, J.; Rogers, G.B.; Morris, P.S.; Smith-Vaughan, H.C.; Nosworthy, E.; Leong, L.E.X.; Smith, R.J.; Weyrich, L.S.; De Haan, J.; Carney, A.S.; et al. The microbiome of otitis media with effusion in Indigenous Australian children. *Int. J. Pediatr. Otorhinolaryngol.* **2015**, *79*, 1548–1555. [CrossRef]

71. Fagö-Olsen, H.; Dines, L.M.; Sørensen, C.H.; Jensen, A. The Adenoids but Not the Palatine Tonsils Serve as a Reservoir for Bacteria Associated with Secretory Otitis Media in Small Children. *mSystems* **2019**, *4*, e00169–e00218. [CrossRef] [PubMed]

72. Johnston, J.; Hoggard, M.; Biswas, K.; Astudillo-García, C.; Radcliff, F.J.; Mahadevan, M.; Douglas, R.G. Pathogen reservoir hypothesis investigated by analyses of the adenotonsillar and middle ear microbiota. *Int. J. Pediatr. Otorhinolaryngol.* **2019**, *118*, 103–109. [CrossRef]

73. Ari, O.; Karabudak, S.; Kalcioglu, M.T.; Gunduz, A.Y.; Durmaz, R. The bacteriome of otitis media with effusion: Does it originate from the adenoid? *Int. J. Pediatr. Otorhinolaryngol.* **2019**, *126*, 109624. [CrossRef] [PubMed]

74. Chan, C.L.; Wabnitz, D.; Bardy, J.J.; Bassiouni, A.; Wormald, P.-J.; Vreugde, S.; Psaltis, A.J. The microbiome of otitis media with effusion. *Laryngoscope* **2016**, *126*, 2844–2851. [CrossRef] [PubMed]

75. Boers, S.A.; De Zeeuw, M.; Jansen, R.; Van Der Schroeff, M.P.; Van Rossum, A.M.C.; Hays, J.P.; Verhaegh, S.J.C. Characterization of the nasopharyngeal and middle ear microbiota in gastroesophageal reflux-prone versus gastroesophageal reflux non-prone children. *Eur. J. Clin. Microbiol. Infect. Dis.* **2018**, *37*, 851–857. [CrossRef] [PubMed]

76. Xu, J.; Dai, W.; Liang, Q.; Ren, D. The microbiomes of adenoid and middle ear in children with otitis media with effusion and hypertrophy from a tertiary hospital in China. *Int. J. Pediatr. Otorhinolaryngol.* **2020**, *134*, 110058. [CrossRef]

77. Chan, C.L.; Wabnitz, D.; Bassiouni, A.; Wormald, P.-J.; Vreugde, S.; Psaltis, A.J. Identification of the Bacterial Reservoirs for the Middle Ear Using Phylogenic Analysis. *JAMA Otolaryngol. Neck Surg.* **2017**, *143*, 155–161. [CrossRef]

78. Krueger, A.; Val, S.; Pérez-Losada, M.; Panchapakesan, K.; Devaney, J.; Duah, V.; DeMason, C.; Poley, M.; Rose, M.; Preciado, D.; et al. Relationship of the Middle Ear Effusion Microbiome to Secretory Mucin Production in Pediatric Patients With Chronic Otitis Media. *Pediatr. Infect. Dis. J.* **2017**, *36*, 635–640. [CrossRef] [PubMed]

79. Kolbe, A.R.; Castro-Nallar, E.; Preciado, D.; Pérez-Losada, M. Altered Middle Ear Microbiome in Children With Chronic Otitis Media With Effusion and Respiratory Illnesses. *Front. Microbiol.* **2019**, *9*, 1–10. [CrossRef]

80. Kim, S.K.; Hong, S.J.; Pak, K.H.; Hong, S.M. Analysis of the Microbiome in the Adenoids of Korean Children with Otitis Media with Effusion. *J. Int. Adv. Otol.* **2019**, *15*, 379–385. [CrossRef]

81. Walker, R.E.; Walker, C.G.; Camargo, C.A., Jr.; Bartley, J.; Flint, D.; Thompson, J.M.D.; Mitchell, E.A. Nasal microbial composition and chronic otitis media with effusion: A case-control study. *PLoS ONE* **2019**, *14*, e0212473. [CrossRef] [PubMed]

82. Neeff, M.; Biswas, K.; Hoggard, M.; Taylor, M.W.; Douglas, R.G. Molecular Microbiological Profile of Chronic Suppurative Otitis Media. *J. Clin. Microbiol.* **2016**, *54*, 2538–2546. [CrossRef] [PubMed]

83. Minami, S.B.; Mutai, H.; Suzuki, T.; Horii, A.; Oishi, N.; Wasano, K.; Katsura, M.; Tanaka, F.; Takiguchi, T.; Fujii, M.; et al. Microbiomes of the normal middle ear and ears with chronic otitis media. *Laryngoscope* **2017**, *127*, E371–E377. [CrossRef] [PubMed]

84. Santos-Cortez, R.L.P.; University of Washington Center for Mendelian Genomics; Chiong, C.M.; Reyes-Quintos, M.R.T.; Tantoco, M.L.C.; Wang, X.; Acharya, A.; Abbe, I.; Giese, A.P.J.; Smith, J.D.; et al. Rare A2ML1 variants confer susceptibility to otitis media. *Nat. Genet.* **2015**, *47*, 917–920. [CrossRef]

85. Santos-Cortez, R.L.P.; Hutchinson, D.S.; Ajami, N.J.; Reyes-Quintos, M.R.T.; Tantoco, M.L.C.; Labra, P.J.; Lagrana, S.M.; Pedro, M.; Llanes, E.G.D.V.; Gloria-Cruz, T.L.; et al. Middle ear microbiome differences in indigenous Filipinos with chronic otitis media due to a duplication in the A2ML1 gene. *Infect. Dis. Poverty* **2016**, *5*, 97. [CrossRef]

86. Hill, C.; Guarner, F.; Reid, G.; Gibson, G.R.; Merenstein, D.J.; Pot, B.; Morelli, L.; Canani, R.B.; Flint, H.J.; Salminen, S.; et al. The International Scientific Association for Probiotics and Prebiotics consensus statement on the scope and appropriate use of the term probiotic. *Nat. Rev. Gastroenterol. Hepatol.* **2014**, *11*, 506–514. [CrossRef]

87. Broek, M.F.L.V.D.; De Boeck, I.; Kiekens, F.; Boudewyns, A.; Vanderveken, O.M.; Lebeer, S. Translating Recent Microbiome Insights in Otitis Media into Probiotic Strategies. *Clin. Microbiol. Rev.* **2019**, *32*, 1–33. [CrossRef]

88. Chen, T.Y.; Hendrickx, A.; Stevenson, D.S.; Bird, P.; Walls, T. No evidence from a systematic review for the use of probiotics to prevent otitis media. *Acta Paediatr.* **2020**. [CrossRef] [PubMed]

89. Lappan, R.; Jamieson, S.E.; Peacock, C.S. Reviewing the Pathogenic Potential of the Otitis-Associated Bacteria Alloiococcus otitidis and Turicella otitidis. *Front. Cell Infect. Microbiol.* **2020**, *10*, 51. [CrossRef]

Review

Oral–Gut Microbiota and Arthritis: Is There an Evidence-Based Axis?

Lorenzo Drago [1,*], Gian Vincenzo Zuccotti [2], Carlo Luca Romanò [3,4], Karan Goswami [5], Jorge Hugo Villafañe [6], Roberto Mattina [7] and Javad Parvizi [5]

1   Laboratory of Clinical Microbiology, Department of Biomedical Sciences for Health & Microbiome, Culturomics and Biofilm related infections (MCB) Unit, "Invernizzi" Pediatric Clinical Research Center, University of Milan, 20133 Milan, Italy
2   Department of Pediatrics, V. Buzzi Childrens' Hospital & "Invernizzi" Pediatric Clinical Research Center University of Milan, 20141 Milan, Italy; gianvincenzo.zuccotti@unimi.it
3   Carlo Luca Romanò, Studio Medico Cecca-Romanò, Corso Venezia, 2, 20121 Milano, Italy; carlo.romano@unimi.it
4   Romano Institute, Rruga Ibrahim Rugova, 1, 00100 Tirane, Albania
5   Rothman Institute, Thomas Jefferson University, Philadelphia, PA 89814, USA; karan.goswami@mail.com (K.G.); javadparvizi@gmail.com (J.P.)
6   IRCCS Fondazione Don Carlo Gnocchi, 20141 Milan, Italy; mail@villafane.it
7   Department of Biomedical, Surgical and Dental Science, University of Milan, 20133 Milan, Italy; roberto.mattina@unimi.it
*   Correspondence: lorenzo.drago@unimi.it

Received: 28 July 2019; Accepted: 15 October 2019; Published: 22 October 2019

Abstract: The gut microbiome appears to be a significant contributor to musculoskeletal health and disease. Recently, it has been found that oral microbiota are involved in arthritis pathogenesis. Microbiome composition and its functional implications have been associated with the prevention of bone loss and/or reducing fracture risk. The link between gut–oral microbiota and joint inflammation in animal models of arthritis has been established, and it is now receiving increasing attention in human studies. Recent papers have demonstrated substantial alterations in the gut and oral microbiota in patients with rheumatoid arthritis (RA) and osteoarthritis (OA). These alterations resemble those established in systemic inflammatory conditions (inflammatory bowel disease, spondyloarthritides, and psoriasis), which include decreased microbial diversity and a disturbance of immunoregulatory proporties. An association between abundance of oral *Porphyromonas gingivalis* and intestinal *Prevotella copri* in RA patients compared to healthy controls has been clearly demonstrated. These new findings open important future horizons both for understanding disease pathophysiology and for developing novel biomarkers and treatment strategies. The changes and decreased diversity of oral and gut microbiota seem to play an important role in the etiopathogenesis of RA and OA. However, specific microbial clusters and biomarkers belonging to oral and gut microbiota need to be further investigated to highlight the mechanisms related to alterations in bones and joints inflammatory pathway.

Keywords: microbiota axis; gut microbiota; oral microbiota arthritis; joint inflammation

---

1. Introduction

Homo sapiens is more prokaryotic than eukaryotic, as the bacteria "layed" in the internal mucosae (intestinal tract, reproductive organs, and respiratory tract) and externally in the body (skin and hair) outnumber host cells 10 to 1 [1]. This paradigm shift has been prompted by the advent of high-throughput metagenomic approaches and has definitively changed the way we study human microbial ecosystems and their interactions with the host. Microbes present in these biological systems are deeply integrated in our daily life, and emerging research has sought to decipher this complex inter-kingdom communication network present in our body and immune system. The gastrointestinal (GI) tract has the highest density and variety of microorganisms (more than 100 trillion microbes and approximately 1500 species). Early life host–microbe interactions, especially in the gut, drive the development of immunity and the establishment of a stable complex microbial community, commonly referred to as the commensal microbiota [2,3]. Extensive research has focused on gut microbiota and host immune response effects in the context of protection against pathogenic gut microbes and the pathophysiology of chronic inflammatory/autoimmune disease states [4,5]. For example, it has been reported that in patients with Crohn's disease, there is a relationship between dysbiosis and response to treatment. Hence, microbiota could be a target of the treatment of chronic intestinal diseases [6].

Emerging scientific reports have also highlighted the immunomodulatory effects of gut microbiota on other pathologic conditions, which often involve distant anatomical sites, such as the liver, the brain, the heart and the skeleton [7–9].

Furthermore, several mechanisms and factors have been implicated to explain the role of microbiota in bone and joint health [10]. The gut microbiome is indeed a source of a number of key vitamins, such as cobalamin (B12), biotin (B7), folate, thiamine (B1), pyridoxal phosphate, pantothenic acid (B5), niacin (B3), vitamin K, and tetrahydrofolate, which are particularly important for the health of the musculoskeletal system [11].

Steves et al. highlighted how the gut microbiome can alter the inflammatory state of an individual by influencing both the host metabolic potential and its innate and adaptive immune system [12]. These authors further discussed the role of microbiota diversity on some prevalent age-related disorders, such as osteoporosis, osteoarthritis, gout, rheumatoid arthritis, frailty and sarcopenia.

In the last decade, the alteration of gut microbiota has been reported in rheumatic disease and arthritis, most notably in juvenile idiopathic arthritis (JIA), rheumatoid arthritis (RA), psoriasis, and the related spondyloarthritides (SpA), including ankylosing spondylitis (AS) and reactive arthritis (ReA) [13]. In a similar fashion to inflammatory bowel disease (IBD), it has been suggested that gut bacteria play important role in the etiopathogenesis of these aforementioned conditions.

RA is an autoimmune disorder which occurs when the immune system affects the fluid that nourishes the cartilage and lubricates the joints (synovium) and their soft tissues. Generally, the root causes of arthritis include an increase in inflammatory processes and a decrease of the normal amount of cartilage present at the joint. A correct diet and gut balance can improve these diseases [14]. Indeed, inflammation-reducing foods containing antioxidants, such as fresh fruits, vegetables, or a gluten-poor diet may improve symptoms and disease progression by restoring intestinal microbiota. Findings have provided a model of how genetic and environmental factors, in association, cause autoimmune diseases such as RA. Sakaguchi S. et al. reported that the causal genetic anomaly of *ZAP-70*, a polymorphism of the *MHC* gene, significantly contributes to determining genetic susceptibility to autoimmune arthritis in SKG mice. Furthermore, they demonstrated that the disease initiation requires the interaction of both genetic and environmental factors, in particular the type of microbial colonization.

One of the most common form of arthritis is osteoarthritis (OA). This disease commonly occurs when the protective cartilage on the ends of bones wears down over time by damaging any joint of the hands, knees, hips and spine. OA is characterized by a chronic, low-grade inflammation which is mediated primarily by the innate immune system, making it distinct from that observed in RA. Several dietary factors have been reported to be involved in the pathogenesis of OA. Vitamins, magnesium, and especially amino acids, i.e., little amounts of single amino acids supplementation such as 0.5% (w/w) L-arginine or 1.0% (w/w) L-glutamine, have shown a significant influence on intestinal microbiota, especially the ratio of Firmicutes/Bacteroidetes. Chitosan supplementation can also alter the component of intestinal microbiota, causing a lowering of the ratio Firmicutes/Bacteroidetes, in particular a decreasing of Bacteroidales and an increasing of the Lactobacillales in the feces [15,16].

The alteration of gut microbiota can thus lead to an increased translocation of microbial associated molecular patterns (MAMPs) across the gut endothelium into the systemic circulation. MAMPs include factors such as lipopolysaccharide (LPS), peptidoglycan, and bacterial DNA. These factors can trigger pro-inflammatory pathways by stimulating immune receptors in the resident immune cells of bone, cartilage and synovium [17,18].

RA has long been associated with periodontal disease [19], and recent evidence on the oral microbiome has emphasized its role in the arthritis. Using a metagenomics approach and molecular investigations, common opinion has been formed that each individual carries over 700 species in the oral cavity, and this microbiome is the second largest microbial niche after the gastrointestinal tract [20]. Oral bacteria may penetrate through the gingival pockets and enter into the bloodstream. The translocation of microbiota-derived molecules into systemic circulation is considered one route for the microbiome to mediate arthritis by stimulating specific cytokines (see below).

There is not so much evidence on microbiota association with some musculoskeletal diseases related to age, as RA and OA. However, it seems that these clinical issues are associated with inflammatory changes, which could be specifically related to microbiota changes or be associated with age. Some studies described below on microbiota and arthritis were age-matched, presuming that the shaping of microbiota may have a role in the developing and maintaining these diseases independently by age.

The present review aims to address the most recent findings regarding the oral and gut microbiomes and their relationship with RA and OA, respectively.

2. Oral Microbiota in RA and OA

RA is an autoimmune disease affecting the synovium and cartilage with bony erosion. Recently, the relationship between the oral microbiome and systemic diseases has been explored [21,22]. Sher et al. demonstrated that overall exposure to *Porphyromonas gingivalis* was similar between patients with RA and controls. These authors found an abundance of *Anaeroglobus geminatus* that correlates with the presence of rheumatoid factors, and *Prevotella* and *Leptotrichia* species are the only taxa that have been observed in patients with new-onset RA [22].

Chen et al. showed that RA has a distinct oral microbiome and may be affected by its dynamic variations [23]. In this study comparing the oral microbiome in RA, OA and healthy patients using rRNA gene amplicon sequencing, eight oral bacterial biomarkers (*Prevotella melaninogenica*, *Veillonella dispar*, *Prevotella*, *Neisseria*, *Porphyromonas*, *Veillonella*, *Haemophilus*, *Rothia*, *Streptococcus*, *Actinomyces*, *Granulicatella*, *Leptotrichia*, *Lautropia*, and *Fusobacterium*) were identified to differentiate RA from OA. In addition, the authors found that patients with RA and OA had oral microbiota with higher microbial diversity compared to healthy subjects, indicating that there could be more pathobionts in the oral cavity of patients with RA that are able to negatively influence the outcome of the disease. The most common phyla were Proteobacteria, Firmicutes, Bacteroidetes, Actinobacteria and Fusobacteria. The relative abundance of Proteobacteria in healthy subjects was significantly higher than in patients with RA and OA, and the relative abundance of Firmicutes in patients with OA is significantly higher than those in patients with RA. Table 1a,b reports the different taxa and species of oral and gut microbiota observed in RA and OA, respectively.

Persson et al. previously noted *P. gingivalis* to be directly linked to RA through citrullination and induction of antipeptidyl citrulline antibodies reacting to citrullinated human self-proteins [24]. Interestingly, *P. gingivalis*, which is mainly abundant in the oral microbiome of RA patients, shares 82% homology of α-enolase with human α-enolase. Consequently, human antibodies against bacterial enolase can promote an increase of antibody production. Lundberg et al. [25] showed that the levels of anti-citrullinated human α-enolase antibodies and bacterial α-enolase correlates with the severity of RA. *P. gingivalis* can be also found in the gut, but nothing exists between the *P. gingivalis* oral–gut axis correlation and arthritis, and the presence of this bacteria in the gut is not an inflammatory trigger of RA.

Table 1. Differences of bacterial abundance (taxa) in oral and gut microbiota of rheumatoid arthritis (RA) (**a**) and osteoarthritis (OA) (**b**) patients compared to the healthy controls.

(a)				
Type of Arthritis	**Abundance**	**Oral Microbiota Profile (Taxa)**	**Abundance**	**Gut Microbiota Profile (Taxa)**
RA	Increase ([23])	*Neisseria subfava, Haemophilus parainfuenzae, Veillonella dispar, Prevotella tannerae, Actinobacillus parahaemolyticus, Neisseria, Haemophilus, Prevotella, Veillonella, Fusobacterium, Aggregatibacter, Actinobacillus*	**Decrease** ([26,27])	*Bacteroides, Akkermansia, F.prausnitzii Prevotella, Ruminococcus*
RA	Increase ([24])	*Porphyromonas gingivalis*		
RA	Decrease ([28])	*Haemophilus* spp.	**Decrease** ([28])	*Haemophilus* spp.
RA	Increase ([28])	*Lactobacillus salivarius*	**Increase** ([28])	*Lactobacillus salivarius*
RA	Increase ([29])	*Plaque: Actinomyces meyeri Prevotella nigrescens Treponema socranskii Treponema* spp. *Eubacterium infirmum Prevotella oris Actimomyces massiliensis Catonella* spp.	**Increase** ([30,31])	*Prevotella copri*
RA	Increase ([29])	*Saliva: Prevotella* spp. *Veillonella* spp. *Centipeda* spp. *Solobacterium morei Prevotella pallens Atopobium parvulum Butyrivibrio* spp.	**Decrease** ([30])	*Bacteroides*
RA	Increase ([32])	*P. melaninogenica P. denticola P. histicola, P. nigrescens, P. oulorum P. maculosa Selenomonas noxia S. sputigena Anaeroglobus geminatus Aggregaticbacter actinomycetemcomitans Parvimonas micra Other Gram-negatives*		

Table 1. *Cont.*

(a)				
RA	Decrease ([32])	*Streptococcus* *Rothia aeria* *Kingella oralis* *Haemophilus* *Actinomyces*		
(b)				
OA	Increase ([28])	*Rothia dentocariosa,* *Ruminococcus gnavus,* *Streptococcus,* *Actinomyces,* *Lautropia,* *Rothia,* *Granulicatella,* *Ruminococcus,* *Oribacterium,* *Abiotrophia*	Increase ([33])	*Lactobacillus* spp. *Methanobrevibacter*
OA			Increase ([33])	*C. coccoides,* *C. leptum,* *Clostridium clusters* *XI-I, Roseburia* spp., *Lactobacillus* spp.
OA			Decrease ([33])	*Bacteroides* *Prevotella* spp.

Eriksson et al. [29], by investigating the periodontal health of patients with RA in relation to oral microbiota and inflammatory levels, found that the majority of the patients had moderate or severe periodontitis and a higher production of anti-citrullinated protein antibodies. The microbiota observed in the plaque were different compared to the saliva samples. The relative bacterial abundances in both sites are shown in Table 1a,b.

A very recent study [32] characterized the subgingival microbiome of RA patients and its association with periodontal severity. The authors demonstrated that changes in the oral microbiota, especially of those species associated with periodontal disease, were linked to worse RA. The abundance of *Prevotella* spp. and the reduction of health-associated species (*Streptococcus*, *Rothia*) may cause an increased production of inflammatory mediators including IL17, IL-2, TNF, and IFN-γ.

Microbial oral translocation into the systemic bloodstream is considered a negative pathway to induce a systemic pro-inflammatory trigger. A recent study reported that the systemic diffusion of bacterial lipopolysaccharide (LPS), a cell wall compound of gram-negatives bacteria, positively correlated with joint inflammatory response and the severity of joint degradation [17]. LPS can also be concentrated into the synovial fluids and upregulate specific pro-inflammatory cytokines. These immunological factors can have an important role in the pathogenesis of arthritis, especially in RA [34]. It is thus probable that many other bacterial clusters and biomarkers can be involved in the increasing of those local or systemic inflammatory conditions which lead to joint/cartilage damage and corrosion.

As mentioned, a clear correlation between bacteria and OA can also be seen by studying the profile of the oral cavity. Oral microbiota seem to have a particular value in OA as well as in the differentiation of RA. Despite these challenging results, more in-depth studies are needed to explore the differences in the oral microbiome profiles of patients with OA. Next-generation sequencing may be a useful tool to further investigate how oral bacteria can affect this type of arthritis.

3. Gut Microbiota in RA and OA

The hypothesis that not only oral but also intestinal microbiota can be associated with the development of RA is supported by many data. Zhang et al. published a case-control metagenome-wide association study (MGWAS) of the fecal, dental and salivary microbiomes of a cohort of treatment-naive and treated RA patients. They found that the RA-associated microbiome deviated significantly from healthy controls in all sites [28]. In this study, *Haemophilus* spp. was depleted in individuals with RA at the fecal and oral levels, whereas *Lactobacillus salivarius* was over-represented in individuals with RA at both microbiota sites.

Older patients often manifest more severe diseases, and this appears connected to age-related gut dysbiosis. Alterations in the microbiota provide plausible candidate mechanisms for driving both inflammation and altering the immune response and host metabolism, which in turn may modulate the development of musculoskeletal problems (see the *Prevotella copri* case below). The microbiome is thus a highly plausible target for the modulation of diseases of aging owing to its close relationship with innate and adaptive immune systems. Components of intestinal microbiota can indeed direct key aspects of host immunity, in particular effector T-cell differentiation, which may impact susceptibility to autoimmune diseases and RA in particular [35].

Different studies investigating the etiology of RA have established the involvement of regulatory T (T-reg) cells, which are defective at suppressing IFN-γ and TNF-α production by conventional T cells in the peripheral blood of active RA patients [36,37]. It has been well established that the gut microbiota–immune interaction and homeostasis, via balancing pro- and anti-inflammatory mechanisms, can regulate the differentiation of various T cell types, especially T-reg cells [38]. A clear example is the potential therapeutic effect of SCFAs (short chain fatty acids), which are microbial fermentation products found in the bowel, that have demonstrated a profound influence on T-reg cell differentiation in a variety of experimental models of autoimmunity or inflammatory T-cell-mediated diseases [39,40].

An elegant collagen-induced arthritis mouse model published by Hui et al. demonstrated that butyrate (a functional SCFA) administration inhibited arthritis by suppressing the expression of inflammatory cytokines [41]. The authors suggested that modulation was likely mediated by the differentiation of CD4 T cells towards T-reg cells, which produce anti-inflammatory cytokine IL-10, and thus influenced the function of Th17 cells.

As mentioned, an altered microbiota profile has also been associated with juvenile idiopathic arthritis (JIA). Current evidence indeed suggests that the perturbation of gut microbiota may contribute to the development of JIA [42,43].

It remains difficult, however, to establish a definitive microbial marker or specific enterotypes that are associated with RA. It has been hypothesized that the alteration of single bacterial genus could have direct impact on driving inflammation, as suggested for *Bacteroides*, *Akkermansia* or the anti-inflammatory *Faecalobacterium prausnitzii*, which has been noted to be depleted in RA patients, while *Prevotella* and *Ruminococcus* were more prevalent [26,27]. Increasing *Prevotella copri* and decreasing *Bacteroides* concentrations in the stool have both been associated with new onset, untreated RA in humans [30].

The above studies, although not always homogenous, have directly or indirectly demonstrated that genetic risk may be modulated by alterations in the microbiome and that the presence of particular microbial markers can be predictive of disease. As mentioned, intestinal microbiota are also known to change with age. Many of the clinical issues, including OA, are related to the inflammatory change—either specific to disease or associated with age. OA is indeed considered a degeneration of joint cartilage and the underlying bone which commonly occurs from middle age onward. The precise etiology of OA remains unknown thus far, even if various risk factors have been associated with presence of the disease, including age, sex, obesity, and diet, and local joint injury [44].

Most of these factors are associated with drastic changes in the intestinal microbiota. Age, in fact, shifts the intestinal microbiota with significant differences between younger adults and older people, showing a lower diversity of gut microbiota, a greater proportion of *Bacteroides* spp., and a distinct abundance pattern of *Clostridium* groups [45]. In addition, obesity is associated with phylum-level changes in the microbiota (i.e., ratio of Firmicutes/Bacteroidetes), reduced bacterial diversity, and an altered representation of bacterial genes and metabolic pathways [46]. Finally, diet can shape gut microbiota and consequently change the composition and metabolism of intestinal microbiota as well as impact host immune responses [47].

For these reasons, OA is now considered an induced inflammatory condition in which the role of the microbiome has emerging as one of the most important factors. Several publications have reported a clear demonstration of the link between osteoarthritis and gut microbiota. For instance, animals with a low-grade chronic systemic inflammation due to a high-fat diet have developed osteoarthritis, and others with an increased body weight due to diet have shown an increased progression of osteoarthritis [48,49]. Metcalfe et al. proposed that metabolic endotoxemia (raised LPS blood and synovial concentrations) caused by impaired gastric mucosa and low-grade chronic inflammation, may contribute to the onset and progression of OA in obese patients [50].

Collins et al. also demonstrated that changes in the Mankin score (a histopathological classification of the severity of osteoarthritic cartilage lesions) seen in a rat model of osteoarthritis were correlated with alterations of gut microbiota [33]. The translocation of bacteria or related compounds (i.e., LPS and peptidoglycans) across the gut barrier into the systemic circulation was found to mediate osteoarthritis. Together, *Lactobacillus* species and *Methanobrevibacter* spp. abundance have shown a strong predictive relationship with the Mankin Score ($p < 0.001$).

Huang and colleagues further demonstrated that systemic and synovial concentrations of bacterial LPS were positively correlated with the joint inflammatory response [17]. This study enrolled 25 patients in whom osteophyte score, joint space narrowing, and pain were measured.

Th epolymerase chain reaction (PCR) analyses and next generation sequencing (NGS) of osteoarthritic synovial fluid and synovial tissue have also revealed the presence of bacterial DNA, raising the possibility that live bacteria or bacterial products are present in the joint during disease progression [51,52].

Other studies have delineated the use of specific probiotics to rebalance gut microbiota and reduce the grade of inflammation. Studies in OA animal models have demonstrated that the oral administration of *Lactobacillus casei* with type II collagen and glucosamine as prebiotic reduces pain, cartilage destruction, and lymphocyte infiltration and leads to a reduced expression of numerous pro-inflammatory cytokines and matrix metalloproteinases, as well as an upregulation of anti-inflammatory cytokines IL-10 and IL-4 [53]. The results observed after the oral intake of a combination of *Lactobacillus casei* and *Lactobacillus acidophilus* in a rat model of collagen-induced arthritis seemed even more protective versus those after indomethacin administration, with regard to oxidative stress parameters in synovial effusate and arthritis scores [54]. A very recent study conducted in a rat model with OA demonstrated that a probiotic diet plus chondroitin sulfate administration reduced the expression of the markers of inflammation and collagen degradation [55].

The exact role of gut microbiota's involvement in the pathophysiology of OA remains under investigation; all these aforementioned observations raise the possibility that the microbiome or part of it may mediate the effects and outcome of this highly prevalent and widespread disease.

4. Discussion

The first description of the possible involvement of microbiota in the pathology of arthritis was published in 1970s when rats raised in germ-free conditions developed severe joint inflammation with 100% penetrance in an adjuvant-induced arthritis model, while conventionally raised controls showed only mild disease at a very low incidence [56].

A fine equilibrium between 'peace-keeping' and potentially pro-inflammatory intestinal and oral bacteria is necessary to keep gut immunity in check and prevent a state of dysbiosis, which might lead to local and distant deleterious consequences in the host. A crucial driver of changes in the gut and oral environments is the inflammatory response of the host. Intestinal and oral inflammations in people are associated with an imbalance in the microbiota, the dysbiosis, which is characterized by a reduced diversity of microbes, a reduced abundance of obligate anaerobic bacteria, and an expansion of facultative anaerobic bacteria in the phylum Proteobacteria, mostly members of the family Enterobacteriaceae.

In regards to RA and gut microbiota, single microorganisms such as *P. copri* might correlate with the development of RA. Pianta A. et al. reported massive concentrations of antibodies against *P. copri* in the sera of RA patients [31]. Impressive advances in sequencing technologies, compelling animal data, and mounting human evidence have suggested that gut microbiota indeed play a part in the pathogenesis of diseases such as autoimmune arthritis. The few studies addressing potential links between the gut microbiota and human inflammatory joint disease have identified dysbiotic patterns that may contribute to initiate or to perpetuate the disease. Obviously, age can greatly contribute to the increase of systemic inflammation (inflammaging), and the microbiota shaped by aging can negatively modulate the outcome of joint diseases. However, the gut microbiota of RA patients seem to be more dysbiotic than those of healthy patients, thus confirming their role as independent of age. An indirect demonstration of the role of microbiota is that gut microbiome (the same for the oral) is perturbed in rheumatoid arthritis and partly normalized after RA-specific treatment [28].

Dysbiosis occurring, for instance, in jejunoileal-bypass, used as alternative to bariatric banding, seems to be associated with arthritis. In these patients, studies have reported a bacterial overgrowth and a deposition of resultant immune complexes in the synovium [57]. However, a very comprehensive human model fitting with the gut–joint axis and dysbiosis can be represented by Whipple's disease, in which the presence of a single bacterium, *Tropheryma whipplei*, overgrowth in the small intestine is sufficient for the development of joint inflammation in predisposed individuals. Similar results have been appointed by some authors regarding the high quantity of *Streptococci* in milk as a theoretical cause of RA [58,59].

A strong evidence of the gut–bone axis and its role in arthritis outcomes has been reported in germ free mice studies. It has been evidenced that these animals do not show arthritis; however, the mono-colonization of particular intestinal bacterium is sufficient to induce arthritis. Thus, gut microbiota have been further confirmed to be a cause of relevant immunological triggers occurring in arthritis pathogenesis [8,60].

Periodontal disease also correlates with new-onset RA patients, and many bacterial clusters related to this disease have been faced in different studies [22–24]. Gene sequencing studies have investigated the subgingival microbiome of patients with RA and compared the results of subjects with osteoarthritis and healthy controls with or without periodontitis. In both cases, literature revealed that specific bacteria biomarker abundance may influence the severity of the osteoarthritic disease. Unfortunately, no unique microbial oral cluster has been identified so far.

Only one study [28] has reported results on the simultaneous effect of oral–gut microbiota in RA patients. By collecting fecal, dental and salivary samples in a cohort of RA and healthy donors, this study demonstrated a rate of dysbiosis in the gut and oral microbiomes of RA patients, suggesting an overlap in the abundance and function of species at different body sites that could be partially resolved after RA treatment.

Despite findings which are suggestive of microbiota–bone axis correlation with inflammatory joint disease, research to date remains inconclusive with regard to the final mechanism. We therefore need to identify the priorities for research in order to substantiate and translate these findings. An important and recent review analyzing nine clinical studies [61] compared changes in diversity and taxa present in the microbiome of RA patients with age, gender and weight-matched controls. Despite microbiome diversity being a generic tool to define whether microbial disturbance in the oral or gut environments has occurred, the study of specific bacterial clusters is of great interest to establish the possible etiopathogenetic role of microbiota in arthritis. In RA, a correlation between a pro-inflammatory genotype-HLA related bacteria and some bacterial clusters has been strongly postulated. However, well-defined human studies using NGS and metabolomic approaches are needed to better understand if and when intestinal community composition in patients with joint inflammation differs (in addition to improving therapies) by looking at specific bacterial markers for disease presence and progression. Prospective studies evaluating the microbiome–host relationship are indeed necessary to establish not only the potential etiology but also the effects of immunosuppressive or anti-inflammatory therapies on microbiota. Another final aim will be to establish how the microbiota can influence therapies per se in OA or RA patients and, subsequently, how they may impact the host's well-being. Table 1a,b shows the main taxa abundances in oral and gut microbiota in OA and RA. To date, interesting and exhaustive data have shown that a connection between microbiomes and joint diseases exists in RA. Other diseases, OA in particular, have received little attention so far, despite some promising, suggestive findings. The gut microbiome, indeed, could be the culprit behind arthritis and joint pain for obese people. A recent paper by Schott E.M. et al. [62] demonstrated that changes in the gut microbiomes of the mice coincided with signs of body-wide inflammation, including in their knees, where the authors induced osteoarthritis with a meniscal tear. Compared to lean mice, osteoarthritis progressed much more quickly in the obese mice, with nearly all of their cartilage disappearing within 12 weeks of the tear.

Though studies have specifically investigated the influence of gut microbiota in OA, pre-clinical data and some observational investigations in humans have suggested a potential relationship between the gut and risk factors of OA. The role of some confounding factors (genes, sex, age, diet, living conditions) needs to be better explored to fully understand the role of gut bacterial biomarkers in OA.

Thus, a deeper understanding of the biological complexities of our 'two genomes' (host and microbial) will help elucidate the factors that trigger inflammation and finally bridge the gap in our knowledge regarding the role of gene–environment interactions in other autoimmune and inflammatory processes involved in disease pathogenesis. Next generation sequencing, metatranscriptomic analysis, and metabolomic approaches may provide yet-greater insight and help to further understand these mechanisms.

There is a justified association between oral and gut microbiomes in arthritis, although the current evidence that the microbiome causes arthritis is far from conclusive. Strategic future studies aiming to improve the understanding of the combined role of gut–oral axis in arthritis as well as the use of "microbiome influencers," such as the probiotics, are mandatory.

5. Highlights of Future Perspectives

Boxes 1–5 report Microbiome definition and its involvement in RA and OA as well as the need for further studies.

Box 1. Microbiome definition.

The microbiome is defined as the totality of microorganisms and their genes inhabiting a unique environment; the human microbiome outnumbers human genes by several orders of magnitude.

Box 2. Tools for studying microbiome.

Understanding of the role of microorganisms in modulating health and disease by NGS and metabolomic technologies will be the new era.

Box 3. Microbiome and RA link.

Despite the fact that precise causation of RA has not yet been established, several clinical investigations have demonstrated the role of some microorganisms in RA pathogenesis, independently of age.

J. Clin. Med. **2019**, *8*, 1753

Box 4. Microbiome and OA link.

OA is the most common disorder of the musculoskeletal system. The literature has considered the microbiome and the use of some selected probiotics as a possible future therapeutic approach.

Box 5. Need for further studies.

More studies are needed to assess the role of the microbiome in human arthritis and related diseases in the order to finally elucidate their mechanisms and therapeutic targets.

Author Contributions: L.D. conceived and write the paper; G.V.Z., C.L.R., R.M., J.H.V. revised the paper; K.G. and J.P. revised the English and improved the manuscript.

Funding: This research received no external funding

Conflicts of Interest: The authors declare no conflict of interest

References

1. Qin, J.; Li, R.; Raes, J.; Arumugam, M.; Burgdorf, K.S.; Manichanh, C.; Nielsen, T.; Pons, N.; Levenez, F.; Yamada, T.; et al. A human gut microbial gene catalogue established by metagenomic sequencing. *Nature* **2010**, *464*, 59–65. [CrossRef] [PubMed]

2. Ivanov, I.I.; Honda, K. Intestinal commensal microbes as immune modulators. *Cell Host Microbe* **2012**, *12*, 496–508. [CrossRef] [PubMed]

3. Sommer, F.; Bäckhed, F. The gut microbiota–masters of host development and physiology. *Nat. Rev. Microbiol.* **2013**, *11*, 227–238. [CrossRef] [PubMed]

4. Bäumler, A.J.; Sperandio, V. Interactions between the microbiota and pathogenic bacteria in the gut. *Nature* **2016**, *535*, 85–93. [CrossRef] [PubMed]

5. Kamada, N.; Seo, S.-U.; Chen, G.Y.; Núñez, G. Role of the gut microbiota in immunity and inflammatory disease. *Nat. Rev. Immunol.* **2013**, *13*, 321–335. [CrossRef] [PubMed]

6. Ribaldone, D.G.; Caviglia, G.P.; Abdulle, A.; Pellicano, R.; Ditto, M.C.; Morino, M.; Fusaro, E.; Saracco, G.M.; Bugianesi, E.; Astegiano, M. Adalimumab Therapy Improves Intestinal Dysbiosis in Crohn's Disease. *J. Clin. Med.* **2019**, *8*, 1646. [CrossRef]

7. Schroeder, B.O.; Bäckhed, F. Signals from the gut microbiota to distant organs in physiology and disease. *Nat. Med.* **2016**, *22*, 1079–1089. [CrossRef]

8. Wu, H.-J.; Ivanov, I.I.; Darce, J.; Hattori, K.; Shima, T.; Umesaki, Y.; Littman, D.R.; Benoist, C.; Mathis, D. Gut-residing segmented filamentous bacteria drive autoimmune arthritis via T helper 17 cells. *Immunity* **2010**, *32*, 815–827. [CrossRef]

9. Li, J.-Y.; Chassaing, B.; Tyagi, A.M.; Vaccaro, C.; Luo, T.; Adams, J.; Darby, T.M.; Weitzmann, M.N.; Mulle, J.G.; Gewirtz, A.T.; et al. Sex steroid deficiency-associated bone loss is microbiota dependent and prevented by probiotics. *J. Clin. Investig.* **2016**, *126*, 2049–2063. [CrossRef]

10. Hernandez, C.J.; Guss, J.D.; Luna, M.; Goldring, S.R. Links Between the Microbiome and Bone. *J. Bone Miner. Res.* **2016**, *31*, 1638–1646. [CrossRef]

11. Kau, A.L.; Ahern, P.P.; Griffin, N.W.; Goodman, A.L.; Gordon, J.I. Human nutrition, the gut microbiome and the immune system. *Nature* **2011**, *474*, 327–336. [CrossRef] [PubMed]

12. Steves, C.J.; Bird, S.; Williams, F.M.K.; Spector, T.D. The Microbiome and Musculoskeletal Conditions of Aging: A Review of Evidence for Impact and Potential Therapeutics. *J. Bone Miner. Res. Off. J. Am. Soc. Bone Miner. Res.* **2016**, *31*, 261–269. [CrossRef] [PubMed]

13. De Filippo, C.; Di Paola, M.; Giani, T.; Tirelli, F.; Cimaz, R. Gut microbiota in children and altered profiles in juvenile idiopathic arthritis. *J. Autoimmun.* **2019**, *98*, 1–12. [CrossRef] [PubMed]

14. Li, Y.; Luo, W.; Deng, Z.; Lei, G. Diet-Intestinal Microbiota Axis in Osteoarthritis: A Possible Role. *Mediat. Inflamm.* **2016**, *2016*, 3495173. [CrossRef]

15. Sakaguchi, S.; Tanaka, S.; Tanaka, A.; Ito, Y.; Maeda, S.; Sakaguchi, N.; Hashimoto, M. Thymus, innate immunity and autoimmune arthritis: Interplay of gene andenvironment. *FEBS Lett.* **2011**, *585*, 3633–3639. [CrossRef]

16. Conti, V.; Leone, M.C.; Casato, M.; Nicoli, M.; Granata, G.; Carlesimo, M. High prevalence of gluten sensitivity in a cohort of patients with undifferentiated connective tissue disease. *Eur. Ann. Allergy Clin. Immunol.* **2015**, *47*, 54–57.

17. Huang, Z.Y.; Stabler, T.; Pei, F.X.; Kraus, V.B. Both systemic and local lipopolysaccharide (LPS) burden are associated with knee OA severity and inflammation. *Osteoarthr. Cartil.* **2016**, *24*, 1769–1775. [CrossRef]

18. Hernandez, C.J. The Microbiome and Bone and Joint Disease. *Curr. Rheumatol. Rep.* **2017**, *19*, 77. [CrossRef]

19. Ogrendik, M. Rheumatoid arthritis is linked to oral bacteria: Etiological association. *Mod. Rheumatol.* **2009**, *19*, 453–456. [CrossRef]

20. Diaz, P.I.; Hoare, A.; Hong, B. Subgingival Microbiome Shifs and Community Dynamics in Periodontal Diseases. *J. Calif. Dent. Assoc.* **2016**, *44*, 397–472.

21. Igari, K.; Kudo, T.; Toyofuku, T.; Inoue, Y.; Iwai, T. Association between periodontitis and the development of systemic diseases. *Oral Biol. Dent.* **2014**, *2*, 4. [CrossRef]

22. Scher, J.U.; Ubeda, C.; Equinda, M.; Khanin, R.; Buischi, Y.; Viale, A.; Lipuma, L.; Attur, M.; Pillinger, M.H.; Weissmann, G.; et al. Periodontal disease and the oral microbiota in new-onset rheumatoid arthritis. *Arthritis Rheum.* **2012**, *64*, 3083–3094. [CrossRef] [PubMed]

23. Chen, B.; Zhao, Y.; Li, S.; Yang, L.; Wang, H.; Wang, T.; Shi, B.; Gai, Z.; Heng, X.; Zhang, C.; et al. Variations in oral microbiome profiles in rheumatoid arthritis and osteoarthritis with potential biomarkers for arthritis screening. *Sci. Rep.* **2018**, *8*, 17126. [CrossRef] [PubMed]

24. Persson, G.R. Rheumatoid arthritis and periodontitis—Inflammatory and infectious connections. Review of the literature. *J. Oral Microbiol.* **2012**, *4*, 11829. [CrossRef] [PubMed]

25. Lundberg, K.; Kinloch, A.; Fisher, B.A.; Wegner, N.; Wait, R.; Charles, P. Antibodies to citrullinated alpha-enolase peptide 1 are specific for rheumatoid arthritis and cross-react with bacterial enolase. *Arthritis Rheum.* **2008**, *58*, 3009–3019. [CrossRef] [PubMed]

26. Wells, C.L.; van de Westerlo, E.M.; Jechorek, R.P.; Feltis, B.A.; Wilkins, T.D.; Erlandsen, S.L. Bacteroides fragilis enterotoxin modulates epithelial permeability and bacterial internalization by HT-29 enterocytes. *Gastroenterology* **1996**, *110*, 1429–1437. [CrossRef]

27. Derrien, M.; Vaughan, E.E.; Plugge, C.M.; de Vos, W.M. Akkermansia muciniphila gen. nov., sp. nov., a human intestinal mucin-degrading bacterium. *Int. J. Syst. Evol. Microbiol.* **2004**, *54*, 1469–1476. [CrossRef]

28. Zhang, X.; Zhang, D.; Jia, H.; Feng, Q.; Wang, D.; Liang, D.; Wu, X.; Li, J.; Tang, L.; Li, Y.; et al. The oral and gut microbiomes are perturbed in rheumatoid arthritis and partly normalized after treatment. *Nat. Med.* **2015**, *21*, 895–905. [CrossRef]

29. Eriksson, K.; Fei, G.; Lundmark, A.; Benchimol, D.; Lee, L.; Hu, Y.; Kats, A.; Saevarsdottir, S.; Catrina, A.I.; Klinge, B.; et al. Periodontal Health and Oral Microbiota in Patients with Rheumatoid Arthritis. *J. Clin. Med.* **2019**, *8*, 630. [CrossRef]

30. Scher, J.U.; Sczesnak, A.; Longman, R.S.; Segata, N.; Ubeda, C.; Bielski, C.; Rostron, T.; Cerundolo, V.; Pamer, E.G.; Abramson, S.B.; et al. Expansion of intestinal Prevotella copri correlates with enhanced susceptibility to arthritis. *eLife* **2013**, *2*, e01202. [CrossRef]

31. Pianta, A.; Arvikar, S.; Strle, K.; Drouin, E.E.; Wang, Q.; Costello, C.E.; Steere, A.C. Evidence of the Immune Relevance of Prevotella copri, a Gut Microbe, in Patients with Rheumatoid Arthritis. *Arthritis Rheumatol.* **2017**, *69*, 964–975. [CrossRef] [PubMed]

32. Corrêa, J.D.; Fernandes, G.R.; Calderaro, D.C.; Mendonça, S.M.S.; Silva, J.M.; Albiero, M.L.; Cunha, F.Q.; Xiao, E.; Ferreira, G.A.; Teixeira, A.L.; et al. Oral microbial dysbiosis linked to worsened periodontal condition in rheumatoid arthritis patients. *Sci. Rep.* **2019**, *9*, 8379. [CrossRef] [PubMed]

33. Collins, K.H.; Paul, H.A.; Reimer, R.A.; Seerattan, R.A.; Hart, D.A.; Herzog, W. Relationship between inflammation, the gut microbiota, and metabolic osteoarthritis development: Studies in a rat model. *Osteoarthr. Cartil.* **2015**, *23*, 1989–1998. [CrossRef] [PubMed]

34. Pretorius, E.; Akeredolu, O.O.; Soma, P.; Kell, D.B. Major involvement of bacterial components in rheumatoid arthritis and its accompaning oxidative stress, systemic inflammation and hypercoagulability. *Exp. Biol. Med. (Maywood)* **2017**, *242*, 355–373. [CrossRef]

35. McLean, M.H.; Dieguez, D.; Miller, L.M.; Young, H.A. Does the microbiota play a role in the pathogenesis of autoimmune diseases? *Gut* **2015**, *64*, 332–341. [CrossRef]

36. Ehrenstein, M.R.; Evans, J.G.; Singh, A.; Moore, S.; Warnes, G.; Isenberg, D.A.; Mauri, C. Compromised function of regulatory T cells in rheumatoid arthritis and reversal by anti-TNFalpha therapy. *J. Exp. Med.* **2004**, *200*, 277–285. [CrossRef]

37. Flores-Borja, F.; Jury, E.C.; Mauri, C.; Ehrenstein, M.R. Defects in CTLA-4 are associated with abnormal regulatory T cell function in rheumatoid arthritis. *Proc. Natl. Acad. Sci. USA* **2008**, *105*, 19396–19401. [CrossRef]

38. Goodrich, J.K.; Waters, J.L.; Poole, A.C.; Sutter, J.L.; Koren, O.; Blekhman, R.; Beaumont, M.; Van Treuren, W.; Knight, R.; Bell, J.T.; et al. Human genetics shape the gut microbiome. *Cell* **2014**, *159*, 789–799. [CrossRef]

39. Furusawa, Y.; Obata, Y.; Fukuda, S.; Endo, T.A.; Nakato, G.; Takahashi, D.; Nakanishi, Y.; Uetake, C.; Kato, K.; Kato, T.; et al. Commensal microbe-derived butyrate induces the differentiation of colonic regulatory T cells. *Nature* **2013**, *504*, 446–450. [CrossRef]

40. Haghikia, A.; Jörg, S.; Duscha, A.; Berg, J.; Manzel, A.; Waschbisch, A.; Hammer, A.; Lee, D.-H.; May, C.; Wilck, N.; et al. Dietary Fatty Acids Directly Impact Central Nervous System Autoimmunity via the Small Intestine. *Immunity* **2015**, *43*, 817–829. [CrossRef]

41. Hui, W.; Yu, D.; Cao, Z.; Zhao, X. Butyrate inhibit collagen-induced arthritis via Treg/IL-10/Th17 axis. *Int. Immunopharmacol.* **2019**, *68*, 226–233. [CrossRef] [PubMed]

42. Stoll, M.L.; Weiss, P.F.; Weiss, J.E.; Nigrovic, P.A.; Edelheit, B.S.; Bridges, S.L.; Danila, M.I.; Spencer, C.H.; Punaro, M.G.; Schikler, S.L.; et al. Age and fecal microbial strain-specific differences in patients with spondyloarthritis. *Arthritis Res. Ther.* **2018**, *20*, 14. [CrossRef] [PubMed]

43. Stoll, M.L.; Kumar, R.; Lefkowitz, E.J.; Cron, R.Q.; Morrow, C.D.; Barnes, S. Fecal metabolomics in pediatric spondyloarthritis implicate decreased metabolic diversity and altered tryptophan metabolism as pathogenic factors. *Genes Immun.* **2016**, *17*, 400–405. [CrossRef] [PubMed]

44. Malfait, A.M. Osteoarthritis year in review 2015: Biology. *Osteoarthr. Cartil.* **2016**, *24*, 21–26. [CrossRef] [PubMed]

45. O'Toole, P.W.; Jeffery, I.B. Gut microbiota and aging. *Science* **2015**, *350*, 1214–1215. [CrossRef] [PubMed]

46. Turnbaugh, P.J.; Hamady, M.; Yatsunenko, T.; Cantarel, B.L.; Duncan, A.; Ley, R.E.; Sogin, M.L.; Jones, W.J.; Roe, B.A.; Affourtit, J.P.; et al. A core gut microbiome in obese and lean twins. *Nature* **2009**, *457*, 480–484. [CrossRef] [PubMed]

47. Ren, W.; Chen, S.; Yin, J.; Duan, J.; Li, T.; Liu, G.; Feng, Z.; Tan, B.; Yin, Y.; Wu, G. Dietary Arginine Supplementation of Mice Alters the Microbial Population and Activates Intestinal Innate Immunity. *J. Nutr.* **2014**, *144*, 988–995. [CrossRef]

48. Mooney, R.A.; Sampson, E.R.; Lerea, J.; Rosier, R.N.; Zuscik, M.J. High-fat diet accelerates progression of osteoarthritis after meniscal/ligamentous injury. *Arthritis Res. Ther.* **2011**, *13*, R198. [CrossRef]

49. Griffin, T.M.; Huebner, J.L.; Kraus, V.B.; Yan, Z.; Guilak, F. Induction of osteoarthritis and metabolic inflammation by a very high-fat diet in mice: Effects of short-term exercise. *Arthritis Rheum.* **2012**, *64*, 443–453. [CrossRef]

50. Metcalfe, D.; Harte, A.L.; Aletrari, M.O.; Al Daghri, N.M.; Al Disi, D.; Tripathi, G.; McTernan, P.G. Does endotoxaemia contribute to osteoarthritis in obese patients? *Clin. Sci. Lond. Engl.* **2012**, *123*, 627–634. [CrossRef]

51. Siala, M.; Gdoura, R.; Fourati, H.; Rihl, M.; Jaulhac, B.; Younes, M.; Sibilia, J.; Baklouti, S.; Bargaoui, N.; Sellami, S.; et al. Broad-range PCR, cloning and sequencing of the full 16S rRNA gene for detection of bacterial DNA in synovial fluid samples of Tunisian patients with reactive and undifferentiated arthritis. *Arthritis Res. Ther.* **2009**, *11*, R102. [CrossRef] [PubMed]

52. Tarabichi, M.; Shohat, N.; Goswami, K.; Alvand, A.; Silibovsky, R.; Belden, K.; Parvizi, J. Diagnosis of Periprosthetic Joint Infection: The Potential of Next-Generation Sequencing. *J. Bone Jt. Surg. Am.* **2018**, *100*, 147–154. [CrossRef] [PubMed]

53. So, J.-S.; Song, M.-K.; Kwon, H.-K.; Lee, C.-G.; Chae, C.-S.; Sahoo, A.; Jash, A.; Lee, S.H.; Park, Z.Y.; Im, S.-H. Lactobacillus casei enhances type II collagen/glucosamine-mediated suppression of inflammatory responses in experimental osteoarthritis. *Life Sci.* **2011**, *88*, 358–366. [CrossRef] [PubMed]

54. Amdekar, S.; Singh, V.; Kumar, A.; Sharma, P.; Singh, R. Lactobacillus casei and Lactobacillus acidophilus regulate inflammatory pathway and improve antioxidant status in collagen-induced arthritic rats. *J. Interferon Cytokine Res.* **2013**, *33*, 1–8. [CrossRef] [PubMed]

55. Korotkyi, O.H.; Vovk, A.A.; Dranitsina, A.S.; Falalyeyeva, T.M.; Dvorshchenko, K.O.; Fagoonee, S.; Ostapchenko, L.I. The influence of probiotic diet and chondroitin sulfate administration on Ptgs2, Tgfb1 and Col2a1 expression in rat knee cartilage during monoiodoacetate-induced osteoarthritis. *Minerva Med.* **2019**, *110*, 419–424. [CrossRef] [PubMed]

56. Kohashi, O.; Kuwata, J.; Umehara, K.; Uemura, F.; Takahashi, T.; Ozawa, A. Susceptibility to adjuvant-induced arthritis among germfree, specific-pathogen-free, and conventional rats. *Infect. Immun.* **1979**, *26*, 791–794.

57. Ross, C.B.; Scott, H.W.; Pincus, T. Jejunoileal bypass arthritis. *Baillieres Clin. Rheumatol.* **1989**, *3*, 339–355. [CrossRef]

58. Moos, V.; Schneider, T. Changing paradigms in Whipple's disease and infection with Tropheryma whipplei. *Eur. J. Clin. Microbiol. Infect. Dis.* **2011**, *30*, 1151–1158. [CrossRef]

59. Svartz, N. The primary cause of rheumatoid arthritis is an infection—The infectious agent exists in milk. *Acta Med. Scand.* **1972**, *192*, 231–239. [CrossRef]

60. Abdollahi-Roodsaz, S.; Joosten, L.A.; Koenders, M.I.; Devesa, I.; Roelofs, M.F.; Radstake, T.R.; Heuvelmans-Jacobs, M.; Akira, S.; Nicklin, M.J.; Ribeiro-Dias, F.; et al. Stimulation of TLR2 and TLR4 differentially skews the balance of T cells in a mouse model of arthritis. *J. Clin. Investig.* **2008**, *118*, 205–216. [CrossRef]

61. Wells, P.M.; Williams, F.M.K.; Matey-Hernandez, M.L.; Menni, C.; Steves, C.J. RA and the microbiome: Do host genetic factors provide the link? *J. Autoimmun.* **2019**, *99*, 104–115. [CrossRef] [PubMed]

62. Schott, E.M.; Farnsworth, C.W.; Grier, A.; Lillis, J.A.; Soniwala, S.; Dadourian, G.H.; Bell, R.D.; Doolittle, M.L.; Villani, D.A.; Awad, H.; et al. Targeting the gut microbiome to treat the osteoarthritis of obesity. *JCI Insight* **2018**, *3*, e95997. [CrossRef] [PubMed]

Journal of
Clinical Medicine

Review

What Pediatricians Should Know before Studying Gut Microbiota

Lorenzo Drago [1,2,*], Simona Panelli [2], Claudio Bandi [2,3], Gianvincenzo Zuccotti [4], Matteo Perini [2] and Enza D'Auria [4]

1   Department of Biomedical Sciences for Health, Università di Milano, 20133 Milan, Italy
2   Department of Biomedical and Clinical Sciences "L. Sacco", Pediatric Clinical Research Center "Invernizzi", Università di Milano, 20157 Milan, Italy
3   Department of Biosciences, Università di Milano, 20133 Milan, Italy
4   Department of Pediatrics, Children's Hospital Vittore Buzzi, Università di Milan, 20141 Milan, Italy
*   Correspondence: lorenzo.drago@unimi.it

Received: 2 July 2019; Accepted: 9 August 2019; Published: 12 August 2019

Abstract: Billions of microorganisms, or "microbiota", inhabit the gut and affect its homeostasis, influencing, and sometimes causing if altered, a multitude of diseases. The genomes of the microbes that form the gut ecosystem should be summed to the human genome to form the hologenome due to their influence on human physiology; hence the term "microbiome" is commonly used to refer to the genetic make-up and gene–gene interactions of microbes. This review attempts to provide insight into this recently discovered vital organ of the human body, which has yet to be fully explored. We herein discuss the rhythm and shaping of the microbiome at birth and during the first years leading up to adolescence. Furthermore, important issues to consider for conducting a reliable microbiome study including study design, inclusion/exclusion criteria, sample collection, storage, and variability of different sampling methods as well as the basic terminology of molecular approaches, data analysis, and clinical interpretation of results are addressed. This basic knowledge aims to provide the pediatricians with a key tool to avoid data dispersion and pitfalls during child microbiota study.

Keywords: gut microbiota; microbiome; maternal–fetal interface; newborn; child; pediatric disease; dysbiosis

---

1. Introduction

The field of microbiome research is quickly evolving and unravelling. Causal links between distinct microbial consortia, their collective functions, and host pathophysiology during the various stages of life are becoming increasingly clear. Studies of microbiome plasticity, composition, and function based on a distinction of the host phenotypes may lay the foundation for both therapeutic and preventive interventions [1]. Indeed, new practical aspects of microbiome studies will be focused on the personalization of actions as well as on an understanding of the inherent individual variability of microbiomes at different ages, stages of development, conditions, and internal or external influences. These studies will allow the comprehension of physiological features to explain, or predict, human health and disease states. Therefore, clinical studies need to be well designed and the subject/patient phenotype properly selected. Age and many other factors have the potential to strongly influence the results, thus clinical studies on microbiota in children should take into account the differences that naturally occur during growth. Other technical challenges that need to be addressed are linked to properly establishing, harmonizing, and standardizing clinical protocols for sample collection, processing, sequencing, and analysis that also takes into account the "microbiome's age". The issues of diet, environment, host immune system, and genetics as key factors for determining microbiome and microbiota profiles have not been fully resolved yet. All of these influences can impact on the microbiota composition at any age and may sometimes be difficult to harmonize and standardize during clinical investigation.

Clinical and microbiological translation urgently needs to implement the main information on microbiota. This review aims to give a rapid overview of child microbiota in order to guide pediatricians to a better understanding of the field while trying to limit biases and intrinsic pitfalls before the study design and starting any clinical trials. Even if most of the reported literature and data specifically refer to the best studied community, in other words, the one inhabiting the gut, the knowledge discussed in the text, together with more practical aspects and recommendations, can also be adapted to the study of other medically-relevant communities (e.g., in nasal-oral cavities).

2. Basic Knowledge on Gut Microbes

The human body harbors trillions of microbial cells mainly represented by bacteria, but also includes archea, viruses, fungi, and parasites. These communities establish extensive networks of cross-feeding (trophic) interactions, consuming, producing, and exchanging hundreds of metabolites with each other and with their human host, with whom they constitute a unique ecological entity called "holobiont" [2,3]. Their highest density is reached in the intestinal compartment, particularly in the lower segments. Here, bacteria are estimated to reach a number of 10^{14} cells and their density in stool have been calculated in the order of 10^{11} per gram of dry material [4]. Although less-well studied, many other body habitats within healthy individuals are occupied by microbial communities such as the mouth and oral tract, nostrils, skin, vagina. The term 'microbiota' literally means all living organisms within a body-site habitat. More specifically, the term "gut microbiota" indicates the resident intestinal bacterial communities, and from a practical point of view, it is generally investigated, with obvious biases, through the analysis of fecal samples, which are easy and non-invasive to collect. The term 'microbiome' is used instead to refer to the genetic content of these microorganisms. Conventionally, research in the field is mainly focused on bacterial microbiome, but further fascinating results have come from the study of "virome", or the viruses inhabiting the gut, of "mycome", which reveals another intriguing world of gut fungi, and of "parasitome".

New genetic and sequencing technologies have opened the way to the 'metagenomic' approach, which directly analyzes the total microbial genomes contained in a sample, that in turn, allows information to be acquired on the genomic links between function and phylogenetic evolution. Other approaches faced in the field include 'metatranscriptomics', the study of the whole RNA repertoire from a microbial community; 'metaproteomics', the study of the entire protein content from the community; and 'meta-metabolomics', the study of small-molecule metabolites produced through the interaction of diet and microbiome [5–7].

The analysis of the gene coding for the ribosomal 16S rRNA is very useful for studying gut bacteria. 16S rRNA is a component of the prokaryotic ribosome and is coded by a gene spanning about 1500 bp. The 16S rRNA gene is highly conserved between different species of bacteria, but presents nine variable ("V") regions that allow identification at the genus or species level. After amplification of, typically, 2–3 V regions, the obtained sequences are clustered into nearly-identical tags called 'phylotypes' or 'operational taxonomic units' (OTUs). These terms refer to a group of microbes generally through the threshold of sequence homology between their 16S rRNA genes (e.g., ≥98% for a 'species'-level phylotype) [8].

Eukaryotic components of the microbiota (e.g., fungi and protozoans) can be analyzed through homologous ribosomal gene sequences (small-subunit rRNA, SSU rRNA), while viral communities that lack ribosomal genes are investigated through shotgun DNA sequencing, or via primers targeted on conserved sequences in viral families. The above approaches are referred to as culture-independent, while culturomics is a culturing approach that uses multiple culture conditions, combined with the MALDI-TOF mass spectrometry and/or the 16S rRNA sequencing, for the isolation and identification of the largest possible number of bacterial species [9].

The gut hosts taxonomically diverse archaea, bacteria, fungi, and viruses. Studies report at least 22 bacterial phyla in the body, mainly represented (>90%) by *Actinobacteria, Firmicutes, Proteobacteria,* and *Bacteroidetes.* In the gut, *Bacteroidetes* and *Firmicutes* represent the predominant phyla [10–12]. In addition to taxonomic composition, taxonomic diversity also needs to be considered in evaluating the homeostasis of microbiota. In particular, two parameters are routinely employed for this purpose: alpha diversity (within-sample diversity, how many taxa or lineages are present in a sample), and beta diversity (between-sample diversity, to which extent the guts of different subjects or patients share taxa or lineages). Parameters that need to be evaluated when computing these ecological indices are richness (i.e., how many bacterial taxa) and evenness, which also takes into account the relative abundance of taxa, in addition to presence/absence, and compares it between subjects or patients [13].

In this context, measures of species richness (for example, the number of observed species or the Chao1 index, which is an abundance-based estimator of species diversity) and phylogenetic measures (Faith's phylogenetic diversity) are sensitive to the number of sequences per sample, whereas this is true to a much minor extent for metrics that combine richness and evenness (Shannon index).

Statistical and computational analyses still remain the main challenge in microbiome research. Some methods currently used for their power and effect size analysis are based on PERMANOVA, Dirichlet Multinomial, or random forest analysis [14]. Parametric statistical tests (for example, the Student's t-test and ANOVA) as well as measures of correlation including Spearman's rank correlation can be used on the basis of the phenotypes under study and the type of information the researcher wants to capture.

3. The Intestinal Microbiota from Birth Throughout Childhood

Addressing neonatal and early-life microbiota is pivotal as many of the events capable of shaping microbial communities even in adults take place during this phase of life: gestational age at birth, type of delivery, breast vs. formula feeding, weaning, use of antibiotics, etc. [15,16]. When neonatal microbiota begins is still a subject of great debate. The "sterile womb paradigm", in other words, the notion that, under physiological conditions, the human fetal environment is sterile and microbial colonization begins with birth, has been accepted for decades. Recently, with the burst of metagenomic studies, there has been a group of papers that have found traces of a lowly abundant bacterial colonization in the placenta, endometrium, amniotic fluid, and meconium in healthy, full-term pregnancies (see Nature Editorial by C. Willyard, 2018, [17] and references therein). This has led some researchers to date back the seeding of the microbiota to before birth ("in utero colonization hypothesis"). The field is still the subject of much debate, and the results appear in general to be controversial. Recently, several scientists have underlined that, even if it is possible that not all healthy babies are born sterile as previously thought, particular caution is necessary when working on samples bearing a low microbial biomass due to the heavy contamination issues notoriously connaturated with such samples when using molecular approaches based on next-generation sequencing [17]. Other important points that have been raised are the difficulty of maintaining a strict sterility when collecting samples related to the in utero environment within a clinical setting, and the impossibility of using NGS-based techniques to discriminate DNA from viable cells and DNA belonging to dead organisms or derived from translocation from the blood stream [15,17].

The human intestine at birth is an aerobic environment, as such, while the adult gut microbiota is dominated by obligate anaerobes belonging to the *Firmicutes* and *Bacteroidetes* phyla, the neonatal pioneer flora is composed by aerotolerant taxa, mainly belonging to the *Enterobacteriaceae* family (phylum: *Proteobacteria*). In a matter of days, however, these microorganisms will reduce oxygen levels, and the intestinal lumen becomes anaerobic. This allows the colonization by strict anaerobes, dominated by *Bifidobacterium* (phylum: *Actinobacteria*); *Clostridium* (phylum: *Firmicutes*); and *Bacteroides* (phylum: *Bacteroidetes*) [18,19]. During the first months, the diet of the infant is almost exclusively milk, favoring milk oligosaccharide fermenters as the already cited *Bifidobacterium*, represented, at this stage, by many species. Other predominant bacterial taxa are represented by *Enterococcaceae*, *Streptococcaceae*, and *Lactobacillaceae* [15].

A very recent paper [20] addressed the development of gut microbiota in a large cohort of children, comprising cases who seroconverted to islet cell autoantibody positivity, children who developed type 1 diabetes (T1D), and matched controls (healthy). This interesting analysis followed the longitudinal maturation of the microbiome from 3 to 46 months of age and determined the covariates that significantly affected its development. Globally, this study harmonized data by collecting 12,500 stool samples from 903 children in three different European countries and three US states. Breastfeeding and birth mode resulted in being the main factors able to drive gut microbiome during the developmental phase by changing some relevant bacterial clusters. The authors proposed three distinct phases of microbiome progression: a developmental phase (months 3–14), a transitional phase (months 15–30), and a stable phase (≥31 months). The Shannon diversity index changed significantly during the first two phases, unchanging only during the stable phase. This study represents a very nice model of how to harmonize the age of the children with other covariate factors. Figure 1 presents a proposal for pediatricians to use a personalized staging of the enrolled individuals to differentiate relevant microbial clusters and dominating phyla.

A step-by-step workflow for a microbiome study

1. Identify a well-designed target population

2. Clearly define primary endpoints of microbiome studies

3. Choosing timing and frequency of sample collection according to study design

4. Use the right storage and sampling methodology

5. Select appropriate sequencing analysis and/or other suitable methodologies

6. Choose the most appropriate computational method

7. Monitor short-term and long-term safety

Figure 1. The figure represents the seven golden steps that the pediatrician should follow before the enrollment of individuals/patients in the microbiota study.

4. Issues to be Considered for Studying Microbiome in Clinical Studies

Study Design and Patient Selections

Pediatricians should select children cohorts by trying to limit the confounding factors that have the potential of diluting the statistical estimates of the effect sizes of the microbiome. Thus, as an example, when defining disease-specific signatures, the diseased population should be recruited with particular care in choosing patients who display a relatively homogeneous clinical phenotype. The choice of controls is also a challenging question: a good control population includes patients with a clinical phenotype that is a clear contrast from the one under study, while matching other relevant criteria. To reduce the heterogeneity of the cohort, it is indeed mandatory to clearly define inclusion and exclusion criteria by considering the factors affecting microbiota analysis (see below) and matching, accordingly, cases and controls. In this regard, it is crucially important to collect information about potential confounding factors, among which age group, for moderating influences that can artifactually alter results and the outcomes of interest. This is important in order to decrease co-variability and heterogeneity during the enrollment, by increasing the power of the analysis in parallel. The collected information will form part of the "metadata" (covariates) surrounding the sample and will later be used in analyzing the data. To ensure consistency, recording the maximum information about the subjects, sample, and experimental procedures is recommended. Finally, before starting the study protocol, a sample size should be estimated on the basis of the expected effect size, and evaluated by means of a pilot study or based on similar previous studies. Other recent approaches rely on computing the estimated sample sizes by calculating the independent effect sizes on microbiota variation of other factors (covariates) relevant to the phenomenon under study [21].

Table 1 summarizes the key aspects to consider when designing and conducting a microbiome study, lists the possible confounders and pitfalls, and presents practical solutions for risk mitigation.

Table 1. Practical aspects to follow when drawing and studying a Microbiome.

Stages and Pitfalls	Considerations and Practical Solutions
Study question	• Clearly define the aim(s) of the study and the relevant biological question(s) before setting up the study design.
Statistically underpowered studies	• Correctly determine the sample size: consider that enrolling enough participants is important to ensure that the expected effect will be detected. • The sample size can be estimated by means of pilot studies, or from previous similar studies, or alternatively from computational approaches that consider the effect of covariates on the total microbiota variation (see main text).
Selection of subjects: avoiding heterogeneity of the population	• Clearly define inclusion and exclusion criteria: consider that an initial heterogeneity of the population will then dilute the statistical estimates of effect sizes on the microbiome. • The list of exclusion criteria from the National Institutes of Health (NIH) Human Microbiome Project can be relied on with regard to the above-mentioned. • In a "cases vs. controls" study, aimed at detecting microbiota-based markers of a disease, choose "cases" with a care in maintaining a relatively homogeneous clinical phenotype. "Controls", in turn, must have a clinical phenotype in clear contrast, while matching other relevant criteria to avoid confounding factors. • Consider that multiple controls groups that are selected based on various criteria may provide more insights. • Additionally consider that for more generalizable results, independent cohorts may be selected to identify the microbiota signatures ("discovery cohort") and test the results ("validation cohort"). • In longitudinal studies, individuals can be treated as their own controls, by collecting baseline samples before and during/after a treatment.
Confounding factors (lifestyle and clinical factors)	• Be exhaustive in the collection of "metadata" (covariates) surrounding the sample: this will be pivotal later, when analyzing the data. Collect information on possible confounding, mediating, and moderating factors that can either influence the microbiome composition or the outcome of interest.
Timing and frequency of sample collection	• Cross-sectional sampling from patients is appropriate to discover and validate diagnostic microbiome signatures. • Repeated samplings of the same subject (time series or longitudinal sampling) ensure more insights into temporal dynamics and community changes. • Longitudinal sampling should be chosen for monitoring disease severity or response to a treatment. Frequency should be similar between subjects.
Sample collection and storage	• Storage and transit conditions are important variables in microbiome study outcomes as they impact DNA yields and quality. • After collecting samples, freeze immediately. When immediate freezing is not possible, short-term refrigeration (+4 °C) is helpful. An alternative is to use stabilizing solutions. • Long-term storage: currently the norm is −80 °C. • Minimize freezing-thawing cycles. To this aim, it is helpful to aliquot samples before freezing.

Table 1. *Cont.*

Stages and Pitfalls	Considerations and Practical Solutions
Experimental Lab procedures	• Use the same procedures and reagents throughout the study. Document everything and be consistent. If, for example, different batches of an enzyme are used, document it among the metadata. • DNA extraction: This is an important source of variation and bias because of the differential resistance to lysis of microbial cells. Combine chemical and mechanical lysing procedures to capture the most accurate community composition. • Contamination may significantly impact results, especially if working on low-biomass samples. It may derive from laboratory contaminants (e.g., previously produced amplicons), from reagents and commercial kits ("kitome"). It is recommendable to separate pre- and post-PCR areas and to introduce appropriate negative controls in different sample processing steps (e.g., blank extraction control: DNA-free water undergoes DNA extraction and all subsequent experimental procedures; blank PCR control: DNA-free water undergoes PCR and all subsequent procedures). • Selection of 16S primers: Rely on previous studies and consider that different couples of universal 16S primers may be biased toward (or against) certain bacterial taxa, thus giving artefactual over- (or under-representations) of them. For example, the 27F/338R primer sets (targeting the V1–V3 regions) is biased against the amplification of *Bifidobacteria*. Another possible pitfall is given by primer sets poorly resolving specific taxa. • PCR amplification: Low DNA template concentration and high number of PCR cycles introduce biases. To reduce their effects, minimize PCR cycles, use a standard (and relatively high) DNA template concentration, and pool multiple PCR (e.g., triplicates) for each sample. The use of proof-reading DNA polymerases and longer annealing times (to reduce chimera formation) is also recommended.
Sequencing	• Use positive controls to calibrate the sequencing method: (i) pure strains of, e.g., *Escherichia coli* that produce strong PCR bands of a known size; and (ii) a synthetic mock microbial community to ensure that amplification, sequencing, and taxonomic classification workflows have not introduced substantial bias or distortions in the expected microbiome profiles. Consider that, in addition to the DNA extraction and PCR steps, errors can be introduced during library preparation, sequencing, imaging, and data analysis.
Data analysis	• The design and choice of the analyses is strictly connected with the research objectives of the study. • Be consistent with the procedures and software used for analyzing data. Consider that different software versions can behave differently. • Integrate non-microbiome sources of data (e.g., clinical parameters) with microbiome data to answer the biological questions that primed the study. • Consider that microbiota data are high-dimensional in nature, with the total number of variable measurements far exceeding the number of samples. • Incorporate the patient and experimental covariates collected in the "metadata" file of the analysis. Evaluate if some of them act as confounding factors. • Repeat the analyses introducing some changes (e.g., change some parameters or algorithms, include or exclude metadata) and the evaluate reproducibility of results. • The complexity of questions in a translational study makes its useful to test multiple statistical models using several combinations of independent-dependent variables. • If a variable is continuous, using it directly in the model is substantially more informative than using a categorical or binary encoding. • Remember that DNA-based techniques are not able to reveal if the microbes under study are alive or dead. If precise information on this is needed, consider performing meta-transcriptomics.
Risk-benefit assessment	• Studies need to be designed to ensure that short term and long-term reliable data are collected.

5. Major Pre-, Peri-, and Post-Natal Factors Affecting the Child Gut Microbiota

A schematic representation of the factors that are able to affect the dynamics and composition of the intestinal microbiota is given in Figure 2.

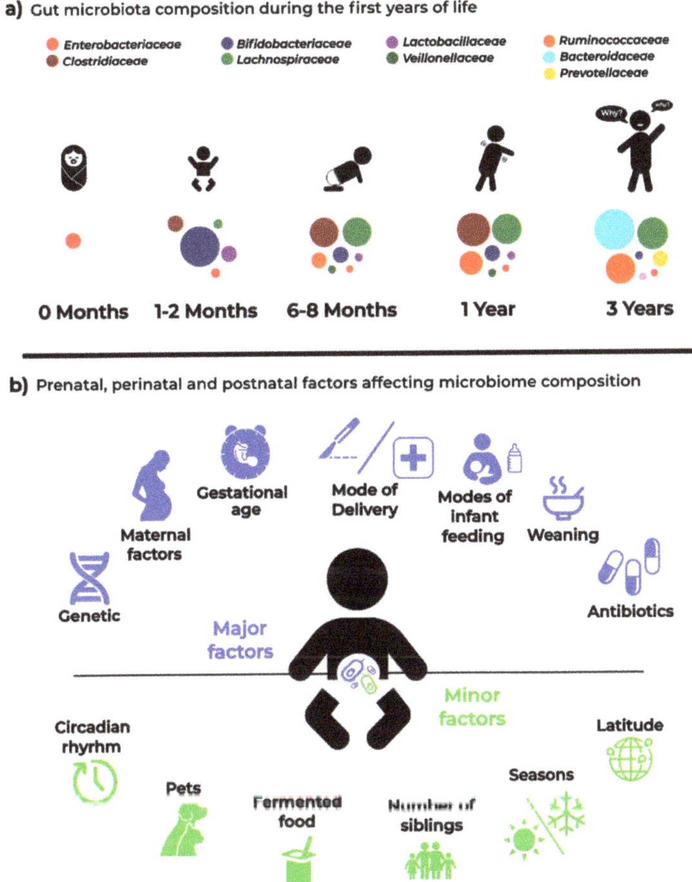

Figure 2. Infant microbiota composition (**a**) and the main "major" and "minor" factors affecting analysis and results in microbiota studies (**b**).

5.1. Maternal Factors Influencing Infant Microbiota

5.1.1. Changes Related to Vertical Transmission of Maternal Metabolites

During gestation, bacteria in the mother's intestine have been shown to drive the future immune maturation of the neonatal gut through the passage of soluble molecules from the placenta in the absence of direct colonization and of the vertical transmission of viable bacterial cells [22,23]. These bacteria are able to induce specific changes in the gut of newborns, creating new microbiota profiles.

5.1.2. Changes Related to Dietary Patterns and Lifestyle

The intestinal microbiota is strongly personalized and influenced by a plethora of environmental and inter-individual variables including body mass index (BMI), exercise frequency, and dietary patterns and habits (which in turn, are strongly related with cultural factors and lifestyle). It has been reported that the infant's fecal microbiota composition is influenced by the BMI and weight gain of the mother during pregnancy [24,25]. In general, the maternal microbial reservoir plays a crucial role in the acquisition and development of early infant microbiota, which in turn is the key to establishing a healthy host–microbiome symbiosis with long-lasting health effects. Therefore, it can be easily understood as to why maternal diet and lifestyle should be monitored and categorized as relevant metadata in infant microbiota studies. In an early phase, after the huge microbial "inoculum" at birth, the infant continues to directly acquire maternal gut strains from different sources (e.g., from skin, mouth, milk) and these are likely to become stable colonizers of the infant gut. Later in life, increasingly important roles are also played by other factors such as shared diet and lifestyle.

5.2. Genetic Factors

There is growing evidence that geographical origin and host genetic makeup influence the acquisition and development of the gut microbiota, with clear associations reported between the host genotype and the relative abundances of different bacterial taxa. For example, Bonder et al. [26] described a single nucleotide polymorphism (SNP) in the LCT locus (coding for human lactase) that is related to varying abundances of *Bifidobacterium*. Goodrich et al. [27], by comparing microbiota across samples belonging to either monozygotic and dizygotic twin pairs, reported a number of microbial taxa whose abundances were strongly influenced by host genetics. Among such taxa, the *Christensellaceae*, considered a microbiome-based marker of obesity and is significantly enriched in individuals with low BMI, resulted in the most highly heritable taxon. Any data related to the genetic hardware of the child should then be noticed.

5.3. Mode of Delivery

At birth, the infant gut communities tend to resemble the maternal vagina or skin microbiota in cases of vaginal or cesarean section (C-section) delivery, respectively [19,28]. Even later, when these "pioneer" foundation populations have been replaced, the birth mode seems to exert significant long-term effects on the structure of the gut microbiota. At 24 months of age, the gut microbial communities of cesarean delivered infants still appear to be less diverse [15]. Even in children as old as seven years, some authors have reported the enduring influence of the mode of delivery, but data are somewhat contrasting regarding this point [19]. Vaginally delivered infants tend to be colonized by *Lactobacillus* and *Prevotella*, while C-section neonates are preferentially colonized by microorganisms from maternal skin, and the hospital staff or environment.

5.4. Mode of Infant Feeding

Breastfed infants receive, from their mothers' milk, a complex mix that will affect the milieu within which their own microbiota will develop. This mix is made up of nutrients, antimicrobial proteins, short chain fatty acids (SCFA), secretory IgA, non-digestible oligosaccharides (HMOs, human milk oligosaccharides, that promote the proliferation of specific gut bacterial taxa in the neonate), and live bacteria, even if previously considered germ-free [15]. The source of the "milk microbiota", which has a transient nature and declines rapidly at weaning, has recently been another subject of debate. At least some of the bacteria is thought to reach the mammary gland through an endogenous route called the enteromammary pathway, which has not been fully elucidated yet. It has also been suggested that mammary skin microbiota can travel via the lymphatic and vascular circulations to the breast ([15,16] and references therein). Gut microbiota differences between breastfed and formula-fed infants are indeed well documented. The former exhibit lower diversity indexes, indicative of a more uniform population where *Bifidobacterium* and *Lactobacillus* dominate. The latter are characterized by more diverse communities, with higher proportions of *Bacteroides*, *Clostridium*, *Streptococcus*, *Veillonella*, *Atopobium*, and *Enterobacteriaceae* [29]. Finally, compositional differences in microbial communities in human milk sampled from different geographical locations have been studied and reported to create strong variability between newborn microbiota [30].

5.5. Gestational Age

While in full-term infants, delivery and feeding mode are reported to represent the major drivers of microbiota development, in preterm (PT) infants (<37 weeks of gestation), the gestational age seems to have the biggest impact on the assembly of gut communities [19,31,32]. PT neonates experience a number of unique challenges in the establishment of their microbiota. Their colonization patterns are characterized by the involvement of peculiar microbial sources, mainly bacteria deriving from the neonatal intensive care unit (NICU) environment [33]. Not rarely, these are strains implicated in nosocomial infections such as *Enterococcus* spp., *Staphylococcus aureus*, *Klebsiella pneumoniae*, *Acinetobacter* spp., *Pesudomonas aeruginosa*, and other *Enterobacteriaceae* [34] with their burden of antibiotic resistance genes. Other relevant features of this peculiar colonization trajectory are its extreme inter-individual variability, and the fact that, across studies, it does not appear to be univocally linked to health outcomes as necrotizing enterocolitis and late-onset sepsis. Instead, the colonization process seems to reflect the co-occurrence of a variety of nosocomial "variables" [35], among which are parenteral nutrition and antibiotic usage (see below). Antibiotics, normally administered to these patients, in turn perturbate the colonization process by killing bacteria acquired during birth and promoting the growth of taxa significantly different from those found in more physiological situations [31]. In conclusion, the PT microbiota appears to be more unstable than that of full-term equivalents and is believed to be associated with a delay in the establishment of an adult-type signature microbiota [16]. All these individuals should be carefully selected and clearly categorized by the clinician before enrollment into the microbiota study.

5.6. Antibiotics

Specific properties of antibiotics, as a mode of action and antimicrobial spectrum, might act as powerful forces for the selection of intestinal bacterial populations, especially if the infant is exposed to antibiotics too early and/or for long periods of time [3,15]. Antibiotics are able to alter the abundances of resident bacteria, significantly impact the growth of otherwise dominant bacterial phyla, and lead to an overall decrease in microbial diversity. A study by Fouhy and colleagues [36] showed that infants exposed to ampicillin and gentamicin shortly after birth harbored higher proportions of *Proteobacteria* and *Actinobacteria*, and the genus *Lactobacillus* for up to four weeks after concluding treatment. Another study reported an attenuation in colonization with *Bifidobacterium* and an increase of *Enterococcus* in subjects receiving oral or intravenous antibiotics during the first four days of life [37].

This variability among individuals suggests caution when including subjects who have been treated with antibiotics [38]. Indeed, the exclusion criteria from the NIH Human Microbiome Project (HMP, dbGAP, see the url https://www.ncbi.nlm.nih.gov/projects/gap/cgi-bin/study.cgi?study_id=phs000228.v4.p1) include the use of systemic antibiotics, antifungals, antivirals, or antiparasitics within six months of sampling. However, this criterion, although optimal, may not be easily applicable with subjects in the pediatric age. For this reason, shorter time windows are often considered. In any case, it is mandatory to accurately document, within the metadata file, any history of antibiotics as well as other medication use.

5.7. Weaning

The transition to more varied, solid food is an important step in the development of the early-life gut microbiota; infants begin to be exposed to a much larger array of substrates and non-digestible carbohydrates that promote the survival and proliferation of more various bacterial taxa. As a consequence, the alpha diversity increases; moreover, *Proteobacteria* and *Actinobacteria* are replaced by *Firmicutes* and *Bacteroidetes* as the dominant phyla, in a more adult-like compositional structure. The cessation of exclusive milk feeding correlates with the decrease of saccharolytic bacteria as *Bifidobacteriaceae* (phylum: *Actinobacteria*). The increased protein intake is thought to be associated with an increase of *Lachnospiraceae* (phylum: *Firmicutes*), while the ingestion of fibers with that of higher levels of *Prevotellaceae* (phylum: *Bacteroidetes*) [39].

In general, the relative abundance of our intestinal microbes is highly influenced by dietary patterns and habits [11], that should therefore be taken into account in clinical studies targeting microbiota.

6. Minor Factors Affecting Gut Microbiota

Various minor factors can affect and modify the gut microbiota, which can occur at any stage of life. Insomnia and circadian rhythm disruption, latitude with time zone shift and intercontinental flights (with the consequent jet lag), household siblings, and companion animals as well as seasonal changes can modify gut microbiota and determine different microbiota profiles with high inter-individual variability to responses to the different factors [40–42]. All of these factors can influence the results and should be carefully considered before starting a clinical study and accurately reported in the metadata to then be considered later in the downstream bioinformatics and statistical analyses. Other similar confounder factors such as bowel movement preparations, evacuants or laxatives, or any microorganism-supplemented food (such as probiotics) can act as deep and long-time gut modifiers, thus a plot-to-plot variation needs to be addressed with nested statistical tests.

7. Sample Collection

Donors/patients to enroll, their genetic or disease phenotypes as well as the expertise of the clinician in methodology used for collecting samples are very relevant in designing a correct study. The number of samples and patients to be enrolled is an intriguing and still hotly debated topic. Sample stability as well as shipping and storage requirements need to be more appropriate and will surely be improved and standardized in the future. Researchers may find some procedures at http://www.microbiome-standards.org or at https://www.hmpdacc.org/resources/metagenomics_sequencing_analysis.php and other papers [43–45].

Concerning the practical aspects, an important question is how often to collect samples because the microbiome ecology is intrinsically dynamic. This largely depends on what question one is trying to answer. If, for gastrointestinal disorders, remarkable changes can be observed between one day and the next (e.g., in times surrounding surgery or in correspondence with periods of activity or remission of the pathology), changes induced by other factors (e.g., diet) often take place on a longer timescale. Collection of multiple samples from the same patient is preferred to allow for better standardization on the basis of the type of patients, centers involved, and statistical power. Whether or not samples collected from the same individual can be pooled before analysis is another topic to be standardized. An important point is that sampling and storage do affect microbiota composition in healthy as well as in diseased subjects. The most widely accepted protocols include immediate homogenization and freezing either with dry ice or in liquid nitrogen, followed by storage at −80 °C. However, this approach is not always practical, particularly for stool samples, or in the case of stool collection from a large scale cohort or remote/rural areas. Whether samples must be immediately frozen (and at what temperature) or whether they can withstand a period of room temperature remains controversial. The above-mentioned studies showed that the effects of short-term storage conditions on the structure and diversity of communities are quite small in general. In particular, storage at −80 °C, −20 °C for a week, or 4 °C for 24 h were found to not significantly affect the ecological indexes of between-sample diversity or the abundance of major taxa [45]. In contrast, the number of freeze–thaw cycles seems to have an effect on the composition of the microbial community, thus it is strongly recommended to aliquot samples at the beginning. Of course, some DNA stabilizers can be used to prolong the stability of samples. In the study of Choo et al. [46] Omnigene Gut and Tris EDTA appeared to show the same performance as storage in an ultrafreezer (−80 °C). In addition to feces, swabs can be an alternative starting material for DNA extractions, especially within hospital settings, even if some studies have shown that the stool swabs of some subjects had limited and not detectable bacterial DNA. A recent study by Christine M. Bassis [47], by comparing stool versus rectal swab samples and their storage conditions, demonstrated minor differences in the bacterial community profiles between the stool and swab from the same subject as well as when samples were stored up to 27 h at +4 °C before freezing at −80 °C. Interestingly, this study also concluded that it was possible to thaw and refreeze samples a limited number of times under particular conditions (i.e., immediately frozen at −20 °C, first thaw cycle, refrozen at −80 °C; immediately frozen at −20 °C, first thaw cycle, refrozen at −20 °C, second thaw cycle and frozen at −80 °C) without strong effects on the community composition. A word of caution is, however, due on this point, as the consensus recommendations are different, as detailed above. Finally, it is to be underlined that as the collection of stool can be difficult from some subjects under certain experimental conditions, swab collection may be useful in such cases, which also has the advantage that they are more easily shipped and handled. A further recent study confirmed that swab samples reliably replicate the stool microbiota bacterial composition when swabs are processed quickly (≤2 days) [48].

Finally, special considerations are needed if addressing peculiar samples such as the newborn's first intestinal discharge (meconium). The debate about "when" the neonatal microbiota begins has been previously mentioned. Recently, several scientists have underlined that, even if it is possible that not all healthy babies are born sterile as previously thought, particular caution is due when working on samples bearing low microbial biomass such as meconia because of the contamination issues connaturated with molecular approaches based on PCR amplification and next-generation sequencing [17,49,50]. The presence of contaminating DNA in laboratory reagents (so-called "kitome") is a serious challenge in these cases; low levels of target bacterial DNA in a sample have been reported to correlate with a high proportion of sequences being attributable to contamination [51,52].

8. Discussion

The Anna Karenina principle, based on Leo Tolstoy's great book and cited in 1878 (*All happy families are alike: each unhappy family is unhappy in its own way*), has been recently translated by Zaneveld et al. [53] as the response to stress against the stability of animal microbiomes. These authors discussed how healthy microbiomes may be quite similar between individuals, but each dysbiotic microbiota is dysbiotic in its own way. The associations between microbiome instability/variability and many confounding factors as well as with diseases, suggest that microbiome may have many and simultaneous multiple faces.

This "stochastic" drift, occurring at any stage of life under stress conditions, can create several phenotypes that need to be known and harmonized when planning a study on microbiota.

Early childhood possesses distinct microbiota tracts compared with later ones, where different clusters and phyla may be differently represented. One common characteristic during this early stage of life is that bacterial richness and diversity increase during growth. Therefore, pediatricians should know that there are several age-related microbiota profiles, and should also be aware of the need to categorize each individual in a defined, monthly range by carefully considering the above-mentioned interference factors.

Several specialties need to be involved in this aim as well as the combination of different knowledge. The "Clinical Microbiota Expert" is not only a new job, but represents a step forward to create competence in this field where clinical microbiologists, clinicians, and bioinformaticians are merged into one. This new job-role will have to create awareness on the study of the "dynamic body" such as the gut microbiota during early age by creating novel models and approaches as well as solutions to solve and interpret the clinical microbiology results. Therefore, translational methodologies to approach a new way of designing clinical trials need to use feasibility and efficacy tools, and a deeper preparation in the field to avoid uncontrolled errors, unsubstantiated results, and excessive costs.

9. Conclusions and Future Perspectives

Next-generation sequencing methodologies still remain expensive and the diagnostic market is offering different solutions, thus a proper, and especially judicious, use of these methods is definitively mandatory. The clinical microbiota expert and pediatricians involved in the field will also have to guide through this jungle by trying to avoid false myths and promises that could be difficult to realize. In the near future, all of these studies and experiences will necessarily lead to a better understanding of the real key phases of microbiome progression from birth throughout childhood.

A final consideration to underline is that the metagenomics community still needs to fully converge toward standardized methods and procedures, leading to an investigation of the sources of variability and bias at each step of the workflow, and to an improved reproducibility and comparability between studies. This is a necessary premise for moving from correlation studies to causation investigations and to answer complex questions in a translational setting.

Author Contributions: L.D. designed and conceived the study; S.P. revised the paper and the technical aspects and checked the literature; C.B. and G.Z. revised the manuscript; M.P. conceived the figures and revised the manuscript; E.D. revised the clinical aspects and the manuscript.

Acknowledgments: The authors wish to thank the Fondazione "Romeo ed Enrica Invernizzi".

Conflicts of Interest: The authors declare no conflicts of interest.

References

1. Zmora, N.; Zilberman-Schapira, G.; Suez, J.; Mor, U.; Dori-Bachash, M.; Bashiardes, S.; Kotler, E.; Zur, M.; Regev-Lehavi, D.; Brik, R.B.Z.; et al. Personalized Gut Mucosal Colonization Resistance to Empiric Probiotics Is Associated with Unique Host and Microbiome Features. *Cell* **2018**, *174*, 1388–1405. [CrossRef] [PubMed]

2. Yatsunenko, T.; Rey, F.E.; Manary, M.J.; Trehan, I.; Dominguez-Bello, M.G.; Contreras, M.; Magris, M.; Hidalgo, G.; Baldassano, R.N.; Anokhin, A.P.; et al. Human gut microbiome viewed across age and geography. *Nature* **2012**, *486*, 222–227. [CrossRef] [PubMed]

3. Rinninella, E.; Raoul, P.; Cintoni, M.; Franceschi, F.; Miggiano, G.A.D.; Gasbarrini, A.; Mele, M.C. What is the Healthy Gut Microbiota Composition? A Changing Ecosystem across Age, Environment, Diet, and Diseases. *Microorganisms* **2019**, *7*, 14. [CrossRef] [PubMed]

4. Lozupone, C.A.; Stombaugh, J.I.; Gordon, J.I.; Jansson, J.K.; Knight, R. Diversity, stability and resilience of the human gut microbiota. *Nature* **2012**, *489*, 220–230. [CrossRef] [PubMed]

5. Williams, S.C. The other microbiome. *Proc. Natl. Acad. Sci. USA* **2013**, *110*, 2682–2684. [CrossRef] [PubMed]

6. Thomas, T.; Gilbert, J.; Meyer, F. Metagenomics—A guide from sampling to data analysis. *Microb. Inform. Exp.* **2012**, *2*, 3. [CrossRef] [PubMed]

7. Turnbaugh, P.J.; Gordon, J.I. An invitation to the marriage of metagenomics and metabolomics. *Cell* **2006**, *134*, 708–713. [CrossRef]

8. Trivedi, B. Microbiome: The surface brigade. *Nature* **2012**, *492*, S60–S61. [CrossRef]

9. Lagier, J.C.; Dubourg, G.; Million, M.; Cadoret, F.; Bilen, M.; Fenollar, F.; Levasseur, A.; Rolain, J.M.; Fournier, P.E.; Raoult, D. Culturing the human microbiota and culturomics. *Nat. Rev. Genet.* **2018**, *16*, 540–550. [CrossRef]

10. Peterson, J.; Garges, S.; Giovanni, M.; McInnes, P.; Wang, L.; Schloss, J.A.; Bonazzi, V.; McEwen, J.E.; Wetterstrand, K.A.; Deal, C.; et al. The NIH human microbiome project. *Genome Res.* **2009**, *19*, 2317–2323.

11. Costello, E.K.; Lauber, C.L.; Hamady, M.; Fierer, N.; Gordon, J.I.; Knight, R. Bacterial Community Variation in Human Body Habitats Across Space and Time. *Science* **2009**, *326*, 1694–1697. [CrossRef] [PubMed]

12. Qin, J.; Li, R.; Raes, J.; Arumugam, M.; Burgdorf, K.S.; Manichanh, C.; Nielsen, T.; Pons, N.; Levenez, F.; Yamada, T.; et al. A human gut microbial gene catalogue established by metagenomic sequencing. *Nature* **2010**, *464*, 59–65. [CrossRef] [PubMed]

13. Human Microbiome Project Consortium. Structure, function and diversity of the healthy human microbiome. *Nature* **2012**, *486*, 207–214. [CrossRef] [PubMed]

14. Kelly, B.J.; Gross, R.; Bittinger, K.; Sherrill-Mix, S.; Lewis, J.D.; Collman, R.G.; Bushman, F.D.; Li, H. Power and sample-size estimation for microbiome studies using pairwise distances and PERMANOVA. *Bioinformatics* **2015**, *31*, 2461–2468. [CrossRef]

15. Arrieta, M.C.; Stiemsma, L.T.; Amenyogbe, N.; Brown, E.M.; Finlay, B. The Intestinal Microbiome in Early Life: Health and Disease. *Front. Immunol.* **2014**, *5*, 427. [CrossRef]

16. Milani, C.; Duranti, S.; Bottacini, F.; Casey, E.; Turroni, F.; Mahony, J.; Belzer, C.; Palacio, S.D.; Montes, S.A.; Mancabelli, L.; et al. The First Microbial Colonizers of the Human Gut: Composition, Activities, and Health Implications of the Infant Gut Microbiota. Microbiol. *Mol. Biol. Rev.* **2017**, *81*, e00036-17. [CrossRef]

17. Willyard, C. Could baby's first bacteria take root before birth? *Nature* **2018**, *553*, 264–266. [CrossRef]

18. Matamoros, S.; Guen, C.G.L.; Le Vacon, F.; Potel, G.; De La Cochetière, M.F. Development of intestinal microbiota in infants and its impact on health. *Trends Microbiol.* **2013**, *21*, 167–173. [CrossRef]

19. Hill, C.J.; Lynch, D.B.; Murphy, K.; Ulaszewska, M.; Jeffery, I.B.; O'Shea, C.A.; Watkins, C.; Dempsey, E.; Mattivi, F.; Touhy, K.; et al. Evolution of gut microbiota composition from birth to 24 weeks in the INFANTMET Cohort. *Microbiome* **2017**, *5*, 4. [CrossRef]

20. Stewart, C.J.; Ajami, N.J.; O'Brien, J.L.; Hutchinson, D.S.; Smith, D.P.; Wong, M.C.; Ross, M.C.; Lloyd, R.E.; Doddapaneni, H.; Metcalf, G.A.; et al. Temporal development of the gut microbiome in early childhood from the TEDDY study. *Nature* **2018**, *562*, 583. [CrossRef]

21. Falony, G.; Joonssens, M.; Vieine-Silva, S.; Wang, J.; Darzi, Y.; Faust, K.; Kurilshikov, A.; Bonder, M.J.; Valles-Colomer, M.; Vandeputte, P.; et al. Population-level analysis of gut microbiota variation. *Science* **2016**, *352*, 560–564. [CrossRef] [PubMed]

22. De Agüero, M.G.; Ganal-Vonarburg, S.C.; Fuhrer, T.; Rupp, S.; Uchimura, Y.; Li, H.; Steinert, A.; Heikenwalder, M.; Hapfelmeier, S.; Sauer, U.; et al. The maternal microbiota drives early postnatal innate immune development. *Science* **2016**, *351*, 1296–1302. [CrossRef] [PubMed]

23. Rakoff-Nahoum, S. Another reason to thank mum: Gestational effects of microbiota metabolites. *Cell Host Microbe* **2016**, *19*, 425–427. [CrossRef] [PubMed]

24. MacPherson, A.J.; De Agüero, M.G.; Ganal-Vonarburg, S.C. How nutrition and the maternal microbiota shape the neonatal immune system. *Nat. Rev. Immunol.* **2017**, *17*, 508–517. [CrossRef] [PubMed]

25. Kearney, J.F.; Patel, P.; Stefanov, E.K.; King, R.G. Natural Antibody Repertoires: Development and Functional Role in Inhibiting Allergic Airway Disease. *Annu. Rev. Immunol.* **2015**, *33*, 475–504. [CrossRef] [PubMed]

26. Bonder, M.J.; Kurilshikov, A.; Tigchelaar, E.F.; Mujagic, Z.; Imhann, F.; Vila, A.V.; Deelen, P.; Vatanen, T.; Schirmer, M.; Smeekens, S.P.; et al. The effect of host genetics on the gut microbiome. *Nat. Genet.* **2016**, *48*, 1407–1412. [CrossRef] [PubMed]

27. Goodrich, J.K.; Waters, J.L.; Poole, A.C.; Sutter, J.L.; Koren, O.; Blekhman, R.; Beaumont, M.; Van Treuren, W.; Knight, R.; Bell, J.T.; et al. Human genetics shape the gut microbiome. *Cell* **2014**, *159*, 789–799. [CrossRef] [PubMed]

28. Dominguez-Bello, M.G.; Costello, E.K.; Contreras, M.; Magris, M.; Hidalgo, G.; Fierer, N.; Knight, R. Delivery mode shapes the acquisition and structure of the initial microbiota across multiple body habitats in newborns. *Proc. Natl. Acad. Sci. USA* **2010**, *107*, 11971–11975. [CrossRef] [PubMed]

29. Fallani, M.; Young, D.; Scott, J.; Norin, E.; Amarri, S.; Adam, R.; Aguilera, M.; Khanna, S.; Gil, A.; Edwards, A.C.; et al. Intestinal Microbiota of 6-week-old Infants Across Europe: Geographic Influence Beyond Delivery Mode, Breast-feeding, and Antibiotics. *J. Pediatr. Gastroenterol. Nutr.* **2010**, *51*, 77–84. [CrossRef]

30. Drago, L.; Toscano, M.; De Grandi, R.; Grossi, E.; Padovani, E.M.; Peroni, D.G. Microbiota network and mathematic microbe mutualism in colostrum and mature milk collected in two different geographic areas: Italy versus Burundi. *ISME J.* **2017**, *11*, 875–884. [CrossRef]

31. Groer, M.W.; Luciano, A.A.; Dishaw, L.J.; Ashmeade, T.L.; Miller, E.; Gilbert, A.J. Development of the preterm infant gut microbiome: A research priority. *Microbiome* **2014**, *2*, 38. [CrossRef] [PubMed]

32. Stewart, C.J.; Embleton, N.D.; Clements, E.; Luna, P.N.; Smith, D.P.; Fofanova, T.Y.; Nelson, A.; Taylor, G.; Orr, C.H.; Petrosino, J.F.; et al. Cesarean or Vaginal Birth Does Not Impact the Longitudinal Development of the Gut Microbiome in a Cohort of Exclusively Preterm Infants. *Front. Microbiol.* **2017**, *8*, 1008. [CrossRef] [PubMed]

33. Brooks, B.; Olm, M.R.; Firek, B.A.; Baker, R.; Thomas, B.C.; Morowitz, M.J.; Banfield, J.F. Strain-resolved analysis of hospital rooms and infants reveals overlap between the human and room microbiome. *Nat. Commun.* **2017**, *8*, 1814. [CrossRef] [PubMed]

34. Hu, H.; Johani, K.; Gosbell, I.; Jacombs, A.; Almatroudi, A.; Whiteley, G.; Deva, A.; Jensen, S.; Vickery, K. Intensive care unit environmental surfaces are contaminated by multidrug-resistant bacteria in biofilms: Combined results of conventional culture, pyrosequencing, scanning electron microscopy, and confocal laser microscopy. *J. Hosp. Infect.* **2015**, *91*, 35–44. [CrossRef] [PubMed]

35. Wandro, S.; Osborne, S.; Enriquez, C.; Bixby, C.; Arrieta, A.; Whiteson, K. The microbiome and metabolome of preterm onfant stool are personalized and not driven by health outcomes, including necrotizing enterocolitis and late-onset sepsis. *Msphere* **2018**, *3*, e00104–e00118. [CrossRef] [PubMed]

36. Fouhy, F.; Guinane, C.M.; Hussey, S.; Wall, R.; Ryan, C.A.; Dempsey, E.M.; Murphy, B.; Ross, R.P.; Fitzgerald, G.F.; Stanton, C.; et al. High-Throughput Sequencing Reveals the Incomplete, Short-Term Recovery of Infant Gut Microbiota following Parenteral Antibiotic Treatment with Ampicillin and Gentamicin. *Antimicrob. Agents Chemother.* **2012**, *56*, 5811–5820. [CrossRef] [PubMed]

37. Tanaka, S.; Kobayashi, T.; Songjinda, P.; Tatejama, A.; Tsubouchi, M.; Kiyohara, C.; Shirakawa, T.; Sonomoto, J.; Nakayama, J. Influence of antibiotic exposure in the early postnatal period in the development of intestinal microbiota. *FEMS Immunol. Med. Microbiol.* **2009**, *56*, 80–87. [CrossRef]

38. Dethlefsen, L.; Relman, D.A. Incomplete recovery and individualized responses of the human distal gut microbiota to repeated antibiotic perturbation. *Proc. Natl. Acad. Sci. USA* **2011**, *108*, 4554–4561. [CrossRef]

39. Koenig, J.E.; Spor, A.; Scalfone, N.; Fricker, A.D.; Stombaugh, J.; Knight, R.; Angenent, L.T.; Ley, R.E. Succession of microbial consortia in the developing infant gut microbiome. *Proc. Natl. Acad. Sci. USA* **2011**, *108*, S4578–S4585. [CrossRef]

40. Thaiss, C.A.; Levy, M.; Korem, T.; Dohnalová, L.; Shapiro, H.; Jaitin, D.A.; David, E.; Winter, D.R.; Gury-BenAri, M.; Tatirovsky, E.; et al. Microbiota Diurnal Rhythmicity Programs Host Transcriptome Oscillations. *Cell* **2016**, *167*, 1495–1510. [CrossRef]

41. Misic, A.M.; Davis, M.F.; Tyldsley, A.S.; Hodkinson, B.P.; Tolomeo, P.; Hu, B.; Nachamkin, I.; Lautenbach, E.; Morris, O.D.; Grice, A.E. The shared microbiota of humans and companion animals as evaluated from *Staphylococcus* carriage sites. *Microbiome* **2015**, *3*, 2. [CrossRef] [PubMed]

42. Davenport, E.R.; Mizrahi-Man, O.; Michelini, K.; Barreiro, L.B.; Ober, C.; Gilad, Y. Seasonal Variation in Human Gut Microbiome Composition. *PLoS ONE* **2014**, *9*, e90731. [CrossRef] [PubMed]

43. Costea, P.I.; Zeller, G.; Sunagawa, S.; Pelletier, E.; Alberti, A.; Levenez, F.; Tramontano, M.; Driessen, M.; Hercog, R.; Jung, F.E.; et al. Towards standards for human fecal sample processing in metagenomic studies. *Nat. Biotechnol.* **2017**, *35*, 1069–1076. [CrossRef] [PubMed]

44. Jordan, S.; Baker, B.; Dunn, A.; Edwards, S.; Ferranti, E.; Mutic, A.D.; Yang, I.; Rodriguez, J. Maternal–Child Microbiome: Specimen Collection, Storage and Implications for Research and Practice. *Nurs. Res.* **2017**, *66*, 175–183. [CrossRef] [PubMed]

45. Tedjo, D.I.; Jonkers, D.M.A.E.; Savelkoul, P.H.; Masclee, A.A.; Van Best, N.; Pierik, M.J.; Penders, J. The Effect of Sampling and Storage on the Fecal Microbiota Composition in Healthy and Diseased Subjects. *PLoS ONE* **2015**, *10*, 0126685. [CrossRef] [PubMed]

46. Choo, J.M.; Leong, L.E.; Rogers, G.B. Sample storage conditions significantly influence faecal microbiome profiles. *Sci. Rep.* **2015**, *5*, 16350. [CrossRef]

47. Bassis, C.M.; Moore, N.M.; Lolans, K.; Seekatz, A.M.; Weinstein, R.A.; Young, V.B.; Hayden, M.K. CDC Prevention Epicenters Program. Comparison of stool versus rectal swab samples and storage conditions on bacterial community profiles. *BMC Microbiol.* **2017**, *17*, 78–85. [CrossRef]

48. Bokulich, N.A.; Maldonado, J.; Kang, D.W.; Krajmalnik-Brown, R.; Caporaso, J.G. Rapidly Processed Stool Swabs Approximate Stool Microbiota Profiles. *mSphere* **2019**, *4*, e00208–e00219. [CrossRef]

49. Lauder, A.P.; Roche, A.M.; Sherrill-Mix, S.; Bailey, A.; Laughlin, A.L.; Bittinger, K.; Leite, R.; Elovitz, M.A.; Parry, S.; Bushman, F.D. Comparison of placenta samples with contamination controls does not provide evidence for a distinct placenta microbiota. *Microbiome* **2016**, *4*, 29. [CrossRef]

50. Perez-Munoz, M.E.; Arrieta, M.C.; Ramer-Tai, A.E.; Walter, J. A critical assessment of the "sterile womb" and "in utero colonization" hypothesis: Implications for research of the pioneer infant microbiome. *Microbiome* **2017**, *5*, 48–67. [CrossRef]

51. Panelli, S.; Schneider, L.; Comandatore, F.; Bandi, C.; Zuccotti, G.V.; D'Auria, E.; Dauria, E. Is there life in the meconium? A challenging, burning question. *Pharmacol. Res.* **2018**, *137*, 148–149. [CrossRef] [PubMed]
52. Ferri, E.; Genco, F.; Gulminetti, R.; Bandi, C.; Novati, S.; Casiraghi, M.; Sambri, V. Plasma Levels of Bacterial DNA in HIV Infection: The Limits of Quantitative Polymerase Chain Reaction. *J. Infect. Dis.* **2010**, *202*, 176–177. [CrossRef] [PubMed]
53. Zaneveld, J.R.; McMinds, R.; Thurber, R.V. Stress and stability: Applying the Anna Karenina principle to animal microbiomes. *Nat. Microbiol.* **2017**, *2*, 17121. [CrossRef] [PubMed]

MDPI
St. Alban-Anlage 66
4052 Basel
Switzerland
Tel. +41 61 683 77 34
Fax +41 61 302 89 18
www.mdpi.com

Journal of Clinical Medicine Editorial Office
E-mail: jcm@mdpi.com
www.mdpi.com/journal/jcm